Studies on the Anatomy and Function of Bone and Joints

Edited by

F. Gaynor Evans

Springer-Verlag   Berlin   Heidelberg   New York 1966

F. Gaynor Evans, Ph. D.
Department of Anatomy
The University of Michigan School of Medicine
Ann Arbor, Michigan, U. S. A.

Symposium organized with the financial assistance of the Council for International Organization of Medical Sciences, an Organization subsidized by the World Health Organization and UNESCO.

ISBN-13: 978-3-642-99911-6   e-ISBN-13: 978-3-642-99909-3
DOI: 10.1007/978-3-642-99909-3

Preface

The various chapters of this monograph were originally presented as papers in a Symposium on Joints and Bones which the editor organized for the VIII International Congress of Anatomists held in Wiesbaden, Germany in August 1965. Each chapter represents original research on the structure and/or function of joints and bones.

Preparing the manuscripts of these papers for publication required more time than originally anticipated and the editor hereby acknowledges his sincere appreciation to the various authors for their help and patience. He also wants to express his special thanks to Mrs. ANTOINETTE CATRON, his editorial assistant, without whose help the task would still be unfinished. The interest and assistance of the staff of Springer-Verlag in the publication of this monograph is also greatly appreciated.

Ann Arbor, Michigan, USA.

February 1966

F. GAYNOR EVANS

Contents

Contributors

ACKERMAN, JAMES L., D.D.S., National Institutes of Health, National Institute of Dental Research, Oral Medicine and Surgery Branch, Oral and Pharyngeal Development Section, Bethesda, Maryland. Present address: Assistant Professor of Orthodontic Research, School of Dentistry, Fairleigh Dickinson University, Teaneck, New Jersey, U.S.A.

ASCENZI, ANTONIO, M.D., Professor of Morbid Anatomy, Institute of Morbid Anatomy, University of Pisa, Pisa, Italy.

BAER, MELVYN J., Ph.D., National Institutes of Health, National Institute of Dental Research, Oral Medicine and Surgery Branch, Oral and Pharyngeal Development Section, Bethesda, Maryland. Present address: Professor of Oral Biology, School of Dentistry, University of Detroit, Detroit, Michigan, U.S.A.

BANG, SEONG, D.D.S., Research Assistant, Department of Anatomy, The University of Michigan, Ann Arbor, Michigan, U.S.A.

BONUCCI, E., M.D., Assistant Professor of Morbid Anatomy, Institute of Morbid Anatomy, University of Pisa, Pisa, Italy.

CHECCUCCI, A., M.D., Researcher of the Italian National Research Council, Institute of Physics, University of Pisa, Pisa, Italy.

DAVIES, D. V., M.D., Professor and Chairman, Department of Anatomy, St. Thomas' Hospital Medical School, London, England.

ENLOW, DONALD H., Ph.D., Associate Professor, Department of Anatomy, The University of Michigan, Ann Arbor, Michigan, U.S.A.

EVANS, F. GAYNOR, Ph.D., Professor, Department of Anatomy, The University of Michigan, Ann Arbor, Michigan, U.S.A.

LISSNER, HERBERT R., M.S., Professor and Chairman, Department of Engineering Mechanics, Wayne State University, Detroit, Michigan, U.S.A. (Deceased).

MACCONAILL, MICHAEL A., M.D., Professor and Chairman, Department of Anatomy, University College, Cork, Ireland.

PALFREY, A. J., M. A., Senior Lecturer, Department of Anatomy, St. Thomas' Hospital Medical School, London, England.

ROBERTS, VERNE L., Ph.D., Associate Professor, Department of Engineering Mechanics and Department of Neurosurgery, Wayne State University, Detroit, Michigan, U.S.A.

RYDELL, NILS, M.D., Docent, Department of Orthopaedic Surgery, University of Gothenburg, Gothenburg, Sweden.

Viidik, Andrus, M.D., Instructor, Department of Anatomy, University of Gothenburg, Gothenburg, Sweden.

Zarek, J. M., Ph.D., Professor and Chairman, Department of Mechanical Engineering, Battersea College of Technology, Proposed University of Surrey, London, England.

Electron Microscopy of Normal Synovial Membrane

D. V. Davies and A. J. Palfrey

The structure of normal synovial membrane as seen under the electron micro-scope is reported for human material by BARLAND, NOVIKOFF and HAMERMAN (1962) and for rabbit synovial membrane by HYDE (1964). It has also been studied in the guinea-pig by WYLLIE, MORE and HAUST (1964), whilst LANGER and HUTH (1960) report its structure in calf, dog and guinea-pig. All these observations have been made on material fixed with osmic acid. The more recently introduced aldehyde fixatives have the great advantage that they do not inactivate all the cell enzymes, but some differences in morphology have been reported in other tissues (SABATINI, BENSCH and BARRNETT, 1962 and 1963; PALFREY and DAVIES, 1966). This paper reports the appearances of normal rabbit synovial membrane after fixation with glutaraldehyde.

Materials and Methods

Synovial membrane was obtained from the knee joint of three male rabbits, aged between three and six months. Under a general anaesthetic 1 ml of fixative was injected into the cavity of the joint; one minute later the cavity was opened and portions of the synovial membrane removed from the region distal to the patella and deep to the ligamentum patellae. The fixative was a 3% solution of glutaraldehyde in a phosphate buffer, which was adjusted to a pH of 7.3 by the dropwise addition of a solution of sodium hydroxide (SABATINI et al., 1963); to this preparation a solution of calcium chloride (0.1 M) was added drop by drop until the precipitate had redissolved (MEEK, personal communication). The speci-mens were cut up in the fixative until no dimension of the block was greater than 1 mm; they remained in fixative for periods of 4 hours (two specimens) or for 1, 3, or 7 days. After fixation the specimens were washed in four changes of a 10% solution of sucrose in the phosphate buffer for ten minutes, and were then treated with a 2% unbuffered solution of osmium tetroxide for 2 hours.

The blocks were dehydrated with ethyl alcohol, passed through propylene oxide and embedded in araldite. Sections for light microscopy were cut at a thickness of 1 μ from the araldite block, and these were stained with Azur II-methylene blue (RICHARDSON, JARETT and FINKE, 1960). Thin sections were cut with a Leitz ultramicrotome at a thickness of about 500 Å and mounted on uncoated copper grids. The mounted sections were stained in a saturated solution of uranyl acetate in absolute methyl alcohol for 1 hour (BARNETT and PALFREY, 1965), before being examined in an A.E.I. electron microscope.

Results

Light Microscopy

Under the light microscope the different blocks taken from similar regions of the same joint in the three different animals vary widely in their appearance. In some the intimal layer is only one cell thick, but in others as many as three or four layers are found. In some blocks this layer lies directly upon adipose tissue, though many blood vessels are seen immediately beneath the intimal layer. In other blocks there is a well marked zone of loose connective tissue amongst which are many prominent blood vessels. This subintimal layer contains a variable number of connective tissue cells, some of which have the form of fibroblasts and fibrocytes, but others are monocytes; in other regions groups of lymphoid cells are found. Many sections show extravasated blood cells, particularly on the surface of the membrane, and less frequently in the substance of the synovial membrane.

Electron Microscopy

When the membrane is examined with the electron microscope the synovial cells are again seen either as a single layer or as a number of layers of cells (Figs. 1—4); the cells which form the surface layer are not in contact with one another but are separated by gaps which vary from as little as 600 Å to as much as 1 μ; in no in-

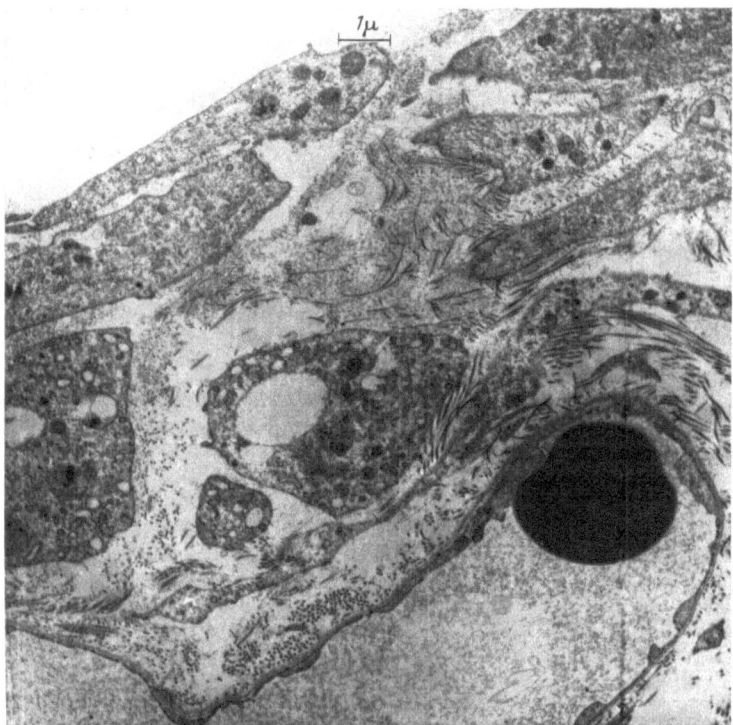

Fig. 1. An electron micrograph of the intimal and subintimal layers of the synovial membrane, showing flattened ill-apposed surface cells with some intervening amorphous material, and a capillary showing fenestrae in the wall. Most cells contain a number of dense bodies. ×7,600

stance is the gap as small as 200 Å, nor is any form of nexus seen between adjacent cells.

These surface cells are not arranged on a basement membrane, but on a zone of material some 0.4 μ in depth (Fig. 4); this contains a few collagen fibres, between 250 and 450 Å in diameter, but which exhibit no periodicity when seen in length. The remainder of this zone is filled with moderately dense granular material which is concentrated in the zone containing few collagen fibres, but extends for a further 0.4 μ amongst the most superficial part of the general collagen framework.

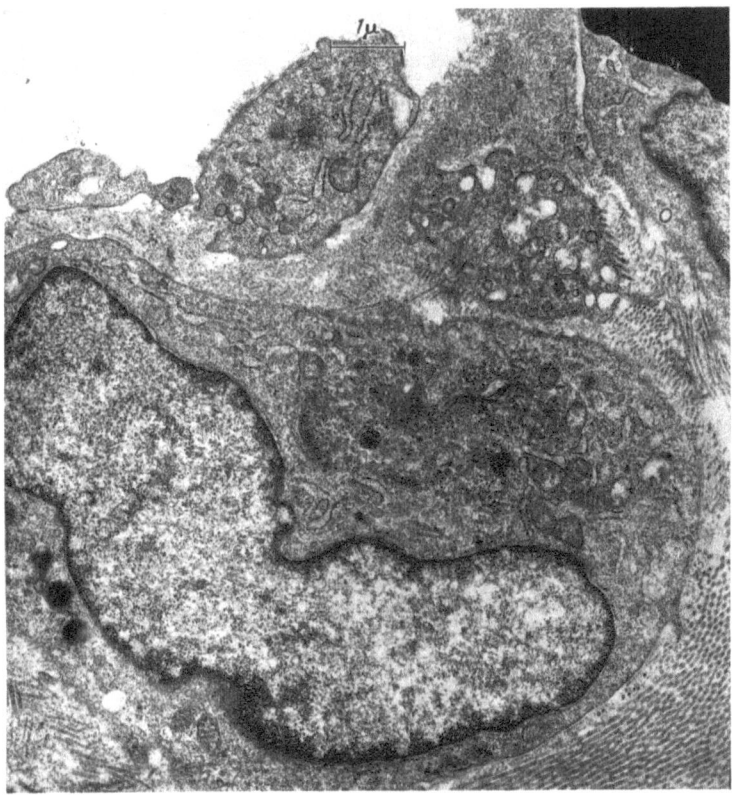

Fig. 2. Intimal layer of the synovial membrane with two cells in which rough surfaced endoplasmic reticulum can be seen. The larger cell shows several mitochondria, Golgi apparatus, and three fat droplets. The amorphous material at the surface is well shown. ×10,500

Surface Cells

The plasma membrane of the surface cells is generally smooth and only occasionally shows filopodia, though such a process is seen in most of these cells (Fig. 3). The filopodia are sinous in form, up to 1 μ in length, and vary in width from 600 to 1000 Å; they are sometimes wider where they change direction, but have bulbous segments which may be terminal. Pinocytotic vesicles, ovoid in form and 900 to 1200 Å in diameter, are found in some cells, and when present are often numerous; they occur with equal frequency on all surfaces of the cells.

The nucleus is somewhat irregular in form, and is limited by the usual double membrane. The outer lamina is clearly seen but is thin (70 Å) while the inner lamina is much thicker — at least 300 Å and may appear even thicker if the chromatin is condensed on its deep surface. The 120 Å interval between the laminae is electron translucent; there are a few areas in which the gap is widened to 600 Å, but these regions show no dense material between the laminae.

The endoplasmic reticulum occurs in short lengths, seldom more than 1 μ, and often branches (Fig. 2). The membranes are poorly visualised and appear to be

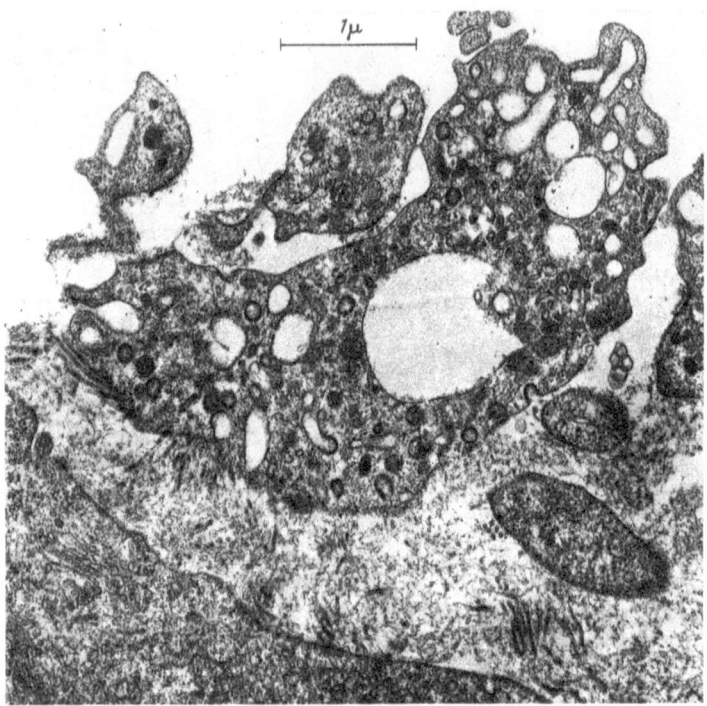

Fig. 3. Surface cell showing processes, an abundance of large vesicles, and pinocytotic vesicles, particularly on its deep surface. ×19,000

40 Å thick; they carry only small numbers of ribonucleoprotein particles, which are about 170 Å in diameter, and are clearly seen. The interval between the members of a membrane pair is usually about 450 Å and is filled with granular material which appears more dense than the surrounding cytoplasm. This interval is widened to as much as 1500 Å in some places; at the ends of the lengths of reticulum the two members of the pair are often continuous with one another to form a vesicle some 1000 Å in diameter. The contents of the vesicle are denser than the material along the length of that section of the reticulum. Many free ribosomes are found in the cytoplasm of some of these cells, but no polyribosomes are identified. In some cells the endoplasmic reticulum is a very prominent feature and the pairs of laminae are arranged concentrically so that the cells are similar in appearance to plasma cells.

Mitochondria are scanty, ovoid in shape, and between 0.4 and 0.8 μ in diameter (Fig. 6). They show clear central areas with sparse irregular cristae which project only about 1000 Å into the cavity; around the cristae there is some dense granular material, but no dense granules were seen.

The Golgi apparatus is well visualised in some cells and may be multiple (Figs. 2 and 9). Stacks of between three and five pairs of membranes are found; each membrane is about 45 Å thick and is well stained. These membrane stacks are surrounded by large numbers of vesicles, particularly at the ends of the membranes;

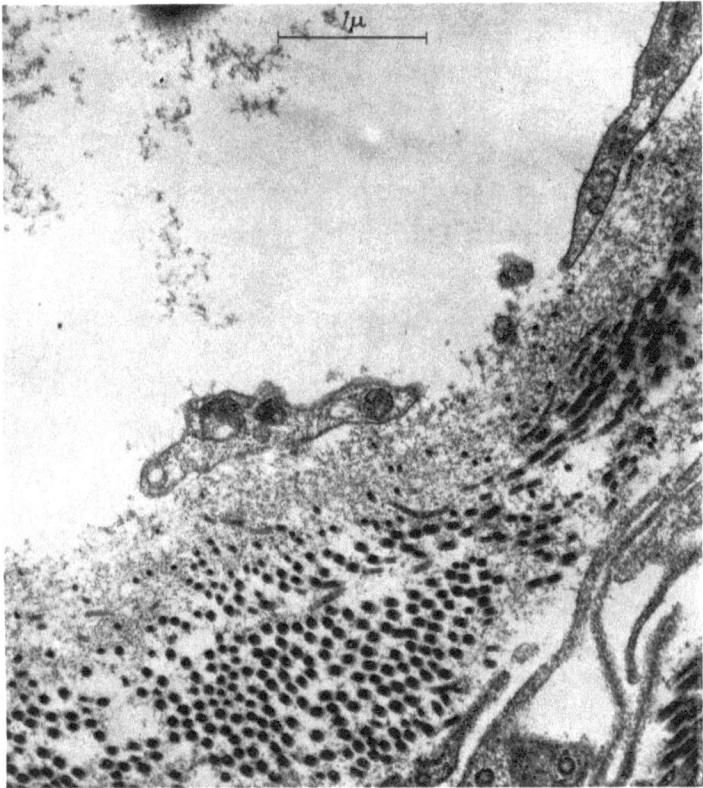

Fig. 4. Surface of synovial membrane showing ill-apposed cells and a distinct layer of amorphous material containing an occasional collagen fibre. ×19,000

the vesicles are between 400 and 800 Å in diameter, and have either a clear central part or may be filled with dense granular material.

Larger vesicles are also found in some of these cells (Fig. 3). They are either circular or ovoid in shape and vary between 0.1 and 1 μ in diameter; each vesicle is limited by a single dense membrane which is 90 Å thick. Their contents are usually electron translucent, though there may be a little granular material at their periphery. Some of these vesicles, including some as large as 0.4 μ in diameter, are seen with their limiting membrane in continuity with the plasma membrane as if they were bursting at the surface of the cell.

Many dense bodies are found in the cytoplasm and these vary in diameter from 750 to 1500 Å (Fig. 1); they may be either ovoid or circular. Their substance is generally of uniform density, though some are granular and others have a lighter peripheral cuff; most are limited by a dense membrane 50 Å thick, but in a few no membrane is present.

A few fat droplets can be seen in some synovial cells (Fig. 2), often occurring in groups of two or three; none has a greater diameter than 0.3 μ. No microtubules are found in any of the cells from the surface of the membrane. Some cells show

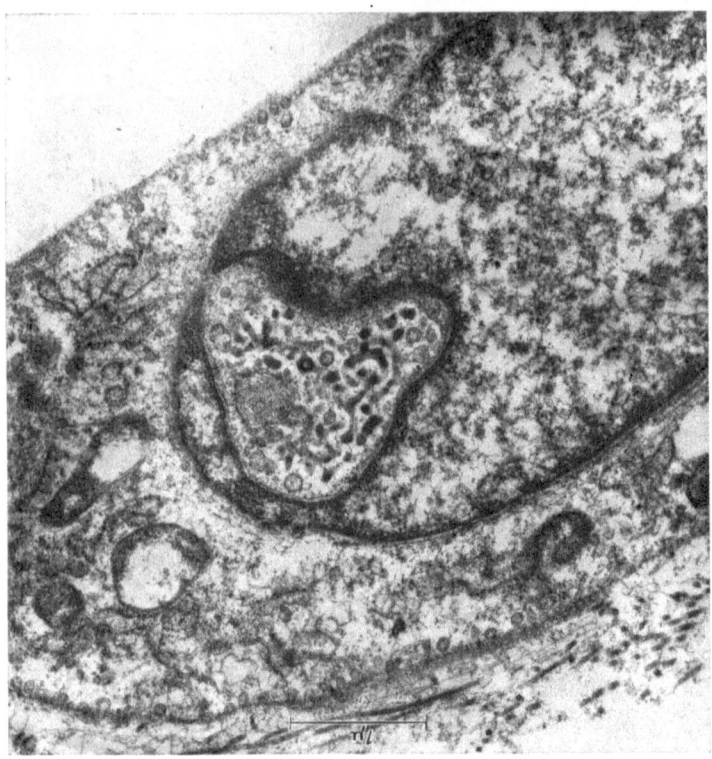

Fig. 5. Surface cell showing an invagination of the nuclear membrane containing a number of vesicles and electron dense strands. ×19,000

many fine filaments in their cytoplasm; these are between 40 and 60 Å in diameter, and are irregularly arranged, though in some places groups of eight to ten filaments can be seen running approximately parallel to one another.

In one surface cell the cytoplasm has formed an invagination into the nucleus (Fig. 5); the ostium in the section studied is 0.1 μ in diameter, but the invagination is dilated to 1.5 μ inside the nucleus. It is limited by the usual two layers of the nuclear membrane and contains some granular material, a number of vesicles between 800 and 1400 Å in diameter, and some electron dense branching strands, each 600 Å wide.

In another cell (Fig. 6) a large vesicle, 1 μ in diameter, is seen limited by a dense 60 Å membrane, and filled with flocculent masses of dense material, which are arranged at the periphery of the vesicle. This mass is continuous with one of the whorled bodies which are a feature of the deeper cells, but are rarely seen in surface cells (see below).

Deeper Layers

In the deeper parts of the intimal layer the cells are scattered amongst bundles of collagen fibres (Fig. 7); these fibres vary between 500 and 1000 Å in diameter, and the characteristic 650 Å banding can be identified. These bundles also contain some unidentified filamentous material, between 50 and 100 Å in diameter, which

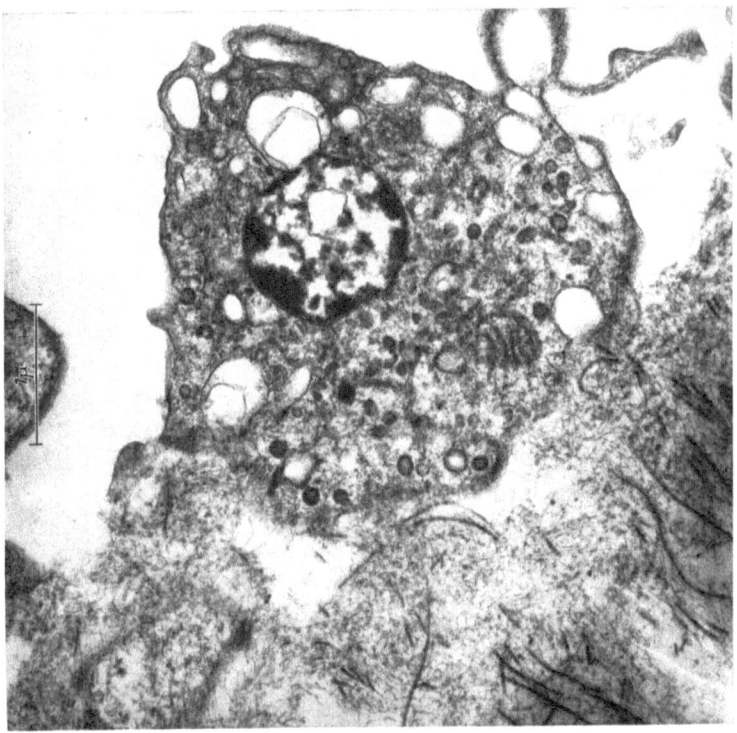

Fig. 6. An intimal cell containing a membrane formation and a large dense body partially filled with electron dense material. Continuous with the dense body there is a large vesicle containing whorled membrane. A single mitochondria with well marked cristae is visible.
×19,000

shows no periodicity. Fat cells are the predominating feature in some parts of the membrane.

In a few sections material can be seen which is probably elastin; this occurs in close proximity to blood vessels, and may be part of the wall of the vessel. It is pensely stained material occurring in bands 1—2 μ in width; these bands are made up by irregularly branching bundles 400 to 800 Å wide and these seem filamentous in structure, though a precise unit of structure cannot be identified.

The cells of the deeper parts of the membrane are similar to those of the surface layer. Their nuclei are rather more irregular; vesicles and dense bodies occur in the same cells; in other cells the Golgi apparatus is well marked, the constituent membranes being well stained. The endoplasmic reticulum is similar in appearance to that seen in the surface cells but dilated portions are more commonly seen, as also is vesicle formation at the ends of the lengths. Fine filamentous material occurs in the cytoplasm more frequently than in the cells from the surface layer.

Microtubules are also a common finding in cells from the deeper part of the membrane, and lengths of up to 1.2 μ can be found (Fig. 8). They are generally straight, though some minor angulations are found in the longer lengths. They are 200 Å in diameter, with a wall 50 Å thick which is much darker than the surrounding cytoplasm; the core is less electron dense, though still darker than the cytoplasm.

Whorled bodies are a common feature in the cytoplasm of the deeper cells; some are found emerging through the plasma membrane and others are seen loose

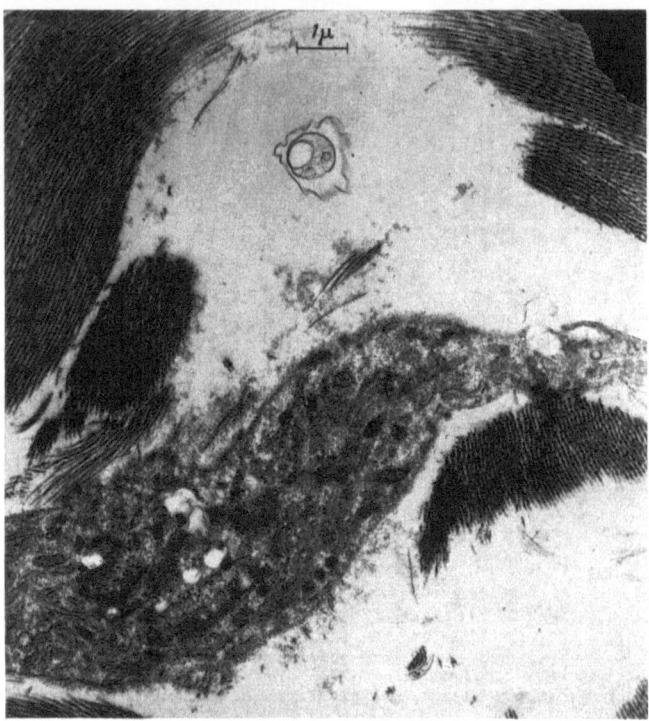

Fig. 7. Subsynovial tissue showing a free whorled membranous body: at the surface of the cell another whorled body is in process of being extruded from a vesicle. ×7,600

in the extracellular space (Figs. 5, 6, and 7). They are rare in the cells of the surface layer. Within the cell they vary between 0.1 and 0.4 μ in diameter, but outside the cells they are larger — between 0.5 and 1 μ in diameter. They consist of two or more concentric layers of membrane which are sometimes seen as a continuous whorl. The membrane is in parts well defined and is then electron dense and 100 Å wide; in other regions, some as long as 0.2 μ, the membrane widens to 800 Å and is then both less definite and less dense. Some of these whorled bodies occur in association with degenerate mitochondria, though others show no such relation; in such mitochondria fat droplets up to 0.2 μ in diameter are also seen.

A centriole is seen in one cell from the deeper parts of the membrane (Fig. 9). This consists of lengths of parallel fibrils, 0.3 μ long and 50 Å in diameter, grouped either in pairs or triplets. These fibrils are surrounded by an ill-defined collection of electron dense granular material about 0.4 μ in diameter; around this there is a zone of clear cytoplasm, and then at a distance of about 0.8 μ from the centriole there is a concentration of Golgi vesicles and membranes.

Another cell from this part of the membrane exhibits a flagellum projecting amongst the collagen fibres (Fig. 10). This is 0.6 μ in length, 0.2 μ wide at its base, narrowing to 0.15 μ near its extremity. The parallel tubules grouped around the

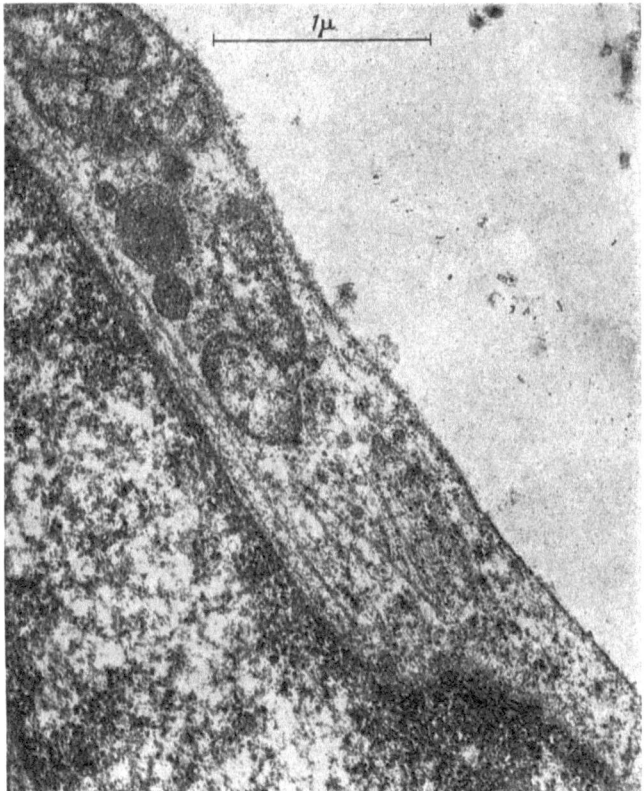

Fig. 8. A cell in the subsynovial tissue showing microtubules. ×29,000

periphery are visualised in pairs and have a wall 60 Å thick, with a cavity about 80 Å wide. The core of the flagellum is 700 Å wide and is composed of clear cytoplasm.

The basal body is also included in the plane of the section; this is 0.25 μ wide and extends 0.4 μ beneath the plasma membrane. Its structure is similar to that of the flagellum, except that it consists of much denser material; the central core is here 850 Å in width. On one side a lateral projection from the basal filaments extends some 800 Å into the cytoplasm, and is about 600 Å wide. In the surrounding cytoplasm there is a concentration of microtubules and also of vesicles and dense bodies, these being about 500 Å in diameter.

The structure of the fat cells is well demonstrated in these preparations (Fig. 11); the bulk of each cell is formed by a single large fat droplet surrounded by a thin rim of featureless cytoplasm. Around some of the fat cells there is a basement membrane-like structure 1000 Å wide, made up of uniform electron-dense material. In the region of the flattened nucleus many smaller fat droplets (500—1000 Å in diameter) are also found in the cytoplasm, but here also other organelles were sparse. In the fat the imperfections of the edge of the knife are always obvious, but chatter is commonly absent; lighter streaks in the fat are seen and these form

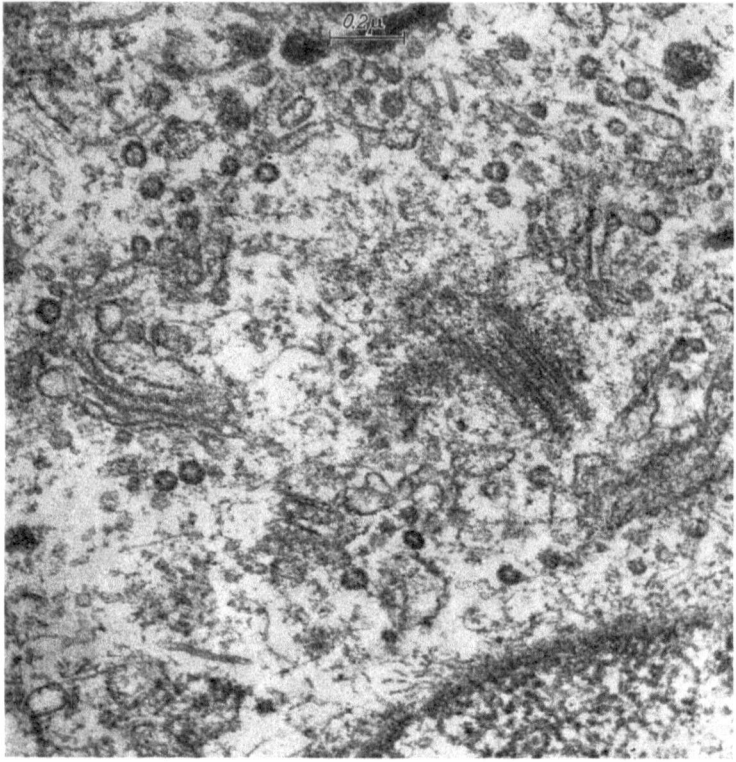

Fig. 9. Part of a subsynovial cell showing a centriole in the concavity of a large semi-circular Golgi apparatus; the edge of the nucleus is seen in the lower right corner. ×49,000

a branching pattern. Irregular patches of cytoplasm are also seen within the fat droplet, and in the midst of these are clear areas containing dense profiles, about 400 Å in diameter; these are thought to be parts of the extracellular space, and the dense profiles to be collagen fibres in cross section.

Blood vessels are a common feature of the synovial membrane (Fig. 1). Those near the surface are lined by a single cell in which the nucleus is only sometimes positioned away from the surface of the synovial membrane. The vessels in the deeper parts of the membrane have walls which are several cells thick; some of these cells can be identified as smooth muscle by their characteristic filamentous cytoplasm. Some of the endothelial cells at all levels show the formation of desmosomes; some also show the formation of pinocytotic vesicles, but these are not a universal feature. In some endothelial cells unusual dense bodies are seen; these are 0.25 µ long, but only 0.08 µ wide; they consist of four or five parallel dense laminae, each about 120 Å wide, with intervening lighter bands of equal width.

Discussion

The morphology of synovial cells as seen both by the light and electron microscope is well documented. Apart from LANGER and HUTH (1960) all investigators are agreed that in most places the lining cellular layer is discontinuous. In defence

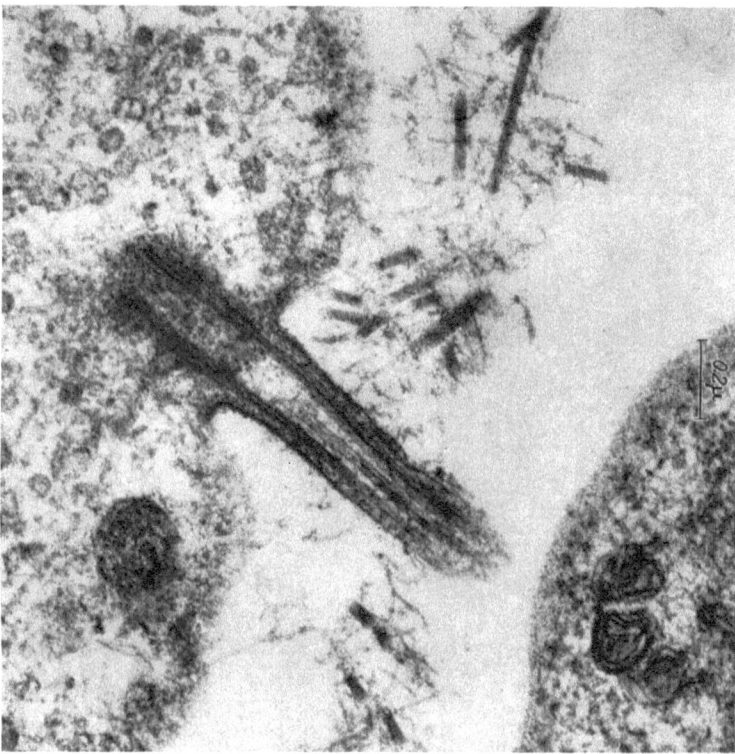

Fig. 10. A well developed flagellum on a subsynovial cell. Note the rootlets arising from the basal body. ×49,000

of the work of LANGER and HUTH (1960) it may be pointed out that the thickness of the intimal layer varies from place to place, and that in pathological conditions such as rheumatoid arthritis where this layer is greatly thickened the cells are closely apposed and display intricate interdigitations. No desmosomes have as yet been reported between these cells either in normal or pathological material.

LANGER and HUTH (1960) describe a basement membrane under the intimal cells. This has not been confirmed by any subsequent investigator. The present study, however, shows that the synovial intimal cells lie in and on amorphous granular material some 0.4 µ deep. Similar amorphous material is described by BARLAND et al. (1962) and by HYDE (1964). This amorphous material seems to be more prominent in glutaraldehyde fixed tissue. This may be due either to its better preservation with this fixative or to variations in its amount in different joints, or perhaps in different species. COULTER (1962) and HYDE (1964) describe collagen fibres at the surface and in contact with the joint space; this was not found by BARLAND et al. (1962) nor is it a feature in the present investigation. The collagen fibres lie more deeply and are everywhere separated from the joint cavity by the amorphous material. A few scattered fibres may be found in this layer.

The nature and significance of the amorphous material presents a problem. Whilst the present investigation provides no clue as to its nature or function, it may be composed of protein, lipoprotein, or mucoprotein, and perhaps provides

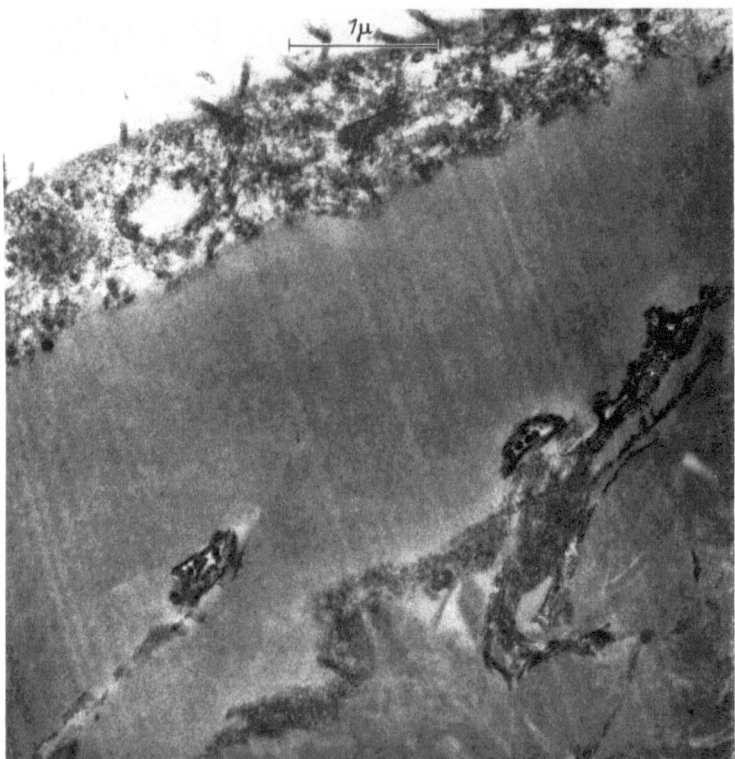

Fig. 11. Part of a fat cell showing connective tissue elements in the fat globule. ×19,000

cohesion between the cells of the intimal layer. FRASER and CATT (1961) have been able to dislodge the cells of the synovial membrane for tissue culture by means of a 0.25% trypsin solution in a phosphate-saline buffer. Hyaluronidase alone is ineffective. FRASER and CATT (1961) do not specify the source of the hyaluronidase though elsewhere they used testicular hyaluronidase, which degrades both hyaluronic acid and chondroitin sulphuric acids. If this was used it is unlikely that the amorphous material is a mucoprotein. There is no information concerning the effect of esterases or phospho-esterases on the cohesion of synovial cells. HAMERMAN and RUSKIN (1959) note a faint metachromatic staining of the matrix and the synovial cells. This is abolished in sections incubated in streptococcal hyaluronidase, which hydrolyses only the hyaluronates amongst the mucopolysaccharides in connective tissue ground substance. On the evidence at present available the amorphous material consists, at least in part, of protein.

The amorphous material may play an important role in the blood-synovial fluid barrier. Whilst in general the composition of synovial fluid is in accord with the view that it is a dialysate of blood serum with the addition of hyaluronic acid, there are certain characteristics in its composition and certain aspects of the permeability of the barrier which remain inadequately explained. Glucose occurs in the synovial fluid in much lower concentration than in serum, whilst thiocyanate passes from the blood into the synovial fluid much less rapidly and in smaller concentration than is to be expected from its diffusibility (BAUER, ROPES and WAINE, 1940).

The results of ENGEL (1940) on the permeability of the blood synovial barrier to dyes need extending and the mechanism involved needs clarifying. ENGEL (1940) shows that, after intravenous injection, acid dyes, presumably as sodium salts, appear in the synovial fluid in a quantity proportionate to their diffusibility in gelatin. Alkaline dyes, presumably as the hydrochlorides, do not appear in the joint perfusate. There are three exceptions to this general rule; Patent blue, Red violet and Methyl orange do not appear in the perfusate despite their acid character and high diffusibility in gelatin. ENGEL (1940) discusses similar experiments by others and notes some discrepancies. Furthermore, less dye diffused into the joint cavity in inflammation and ENGEL (1940) attributes this to some change in the synovial membrane, though again there are discrepancies. ENGEL (1940) suggests that the selective permeability may depend on the lipid solubility of the dye, for alkaline dyes pass into and are retained by the cells. Methyl orange is lipid soluble, but in similar experiments in which he has perfused the peritoneal cavity the dye appears in the perfusate, and some special property is attributed to the synovial barrier. The permeability of the barrier seems to be diminished by sympathectomy (ENGEL, 1941). The composition of the amorphous layer described in the present investigation clearly needs further study, and so does the permeability of the barrier in both health and disease. It would appear that glutaraldehyde is the best fixative so far used for its preservation.

The above problem may be related to the identity of the whorled bodies, seen both in the intracellular vesicles and intercellular spaces in the subintimal layer. Phospholipid is well known to produce whorled formations when fixed.

BARLAND et al. (1962) describe two types of synovial cell. Type A contains numerous vacuoles, many processes at the joint surface, intracellular fibrils, frequent mitochondria, and a prominent Golgi apparatus. Type B is characterised by an abundant rough-walled endoplasmic reticulum, fewer and smaller vacuoles, and mitochondria. CHAPMAN et al. (1962), BALL et al. (1964) and WYLLIE et al. (1964) find some evidence for a similar differentiation of the cell types but mention transitional types. LEVER and FORD (1958), LANGER and HUTH (1960), and COULTER (1962) did not distinguish two types. CHAPMAN et al. (1962) have sought to correlate the morphology of these cells with function. Using iron dextran injected into the joint cavity they showed significant concentrations in the vacuolated cells with lesser amounts in the type B cell which has been supposed to be concerned with the synthesis of hyaluronic acid. A distinction between these two cell types cannot be made out in all regions of the synovial membrane. They may, as already suggested by others, represent phases in the activity of synovial cells, or different regions of the synovial membrane may show different structural characteristics.

The cells of the synovial membrane secrete hyaluronic acid; this has been shown in tissue culture by Vaubel (1933a and b), Kling et al. (1955), Grossfield et al. (1955) and Castor (1957). Castor (1957) points out that these cultures contain mixtures of cell types. Furthermore, no specific unequivocal method for staining of hyaluronate has been evolved and no specific organelle or region of the synovial cells has been identified with the synthesis of this product. The most prominent and highly developed organelle in the synovial cells is the Golgi apparatus. This becomes enlarged and even more prominent when the membrane is irritated as by disease or by injection of materials into the joint cavity (Cochrane et al., 1965). In tissue culture Hedberg and Moritz (1958) have obtained an equally intense production of hyaluronic acid from the fibroblasts of the periarticular tissues as from synovial cells. It would appear that both the hexosamine and glucuronic components of the hyaluronic acid molecule can be synthesised from glucose (Yielding et al., 1957). The prominent Golgi apparatus might be considered a likely site of production in the cell, but an autoradiographic study at electron microscope level is needed. A number of oxidative enzymes, lactic dehydrogenase — D.P.N.H. — diaphorase, D.P.N.H. — diaphorase and succinic dehydrogenase, have been shown to be present in higher concentration in the lining cells of the synovial membrane than in the subintimal tissue of the synovial membrane (Hamerman and Blum, 1959). Enzymes involved in the synthesis of the glucuronic acid have been demonstrated in extracts of the synovial membrane (Thomas et al., 1958). These findings suggest that the synovial cells synthesise hyaluronate.

The present investigation shows for the first time that the cells deep to the surface of the synovium possess microtubules; these have not been seen in the surface cells. Their functions, whether concerned with mobility, forming a transport system, or providing an intracellular skeleton, remain undecided.

The presence of a flagellum on a fibroblast in the subsynovial tissue recalls the work of Sorokin (1962) wherein cilia are described on fibroblasts and smooth muscle cells. A role for the centriolar bodies in ciliogenesis is advanced by Sorokin (1962). Centrioles are present in synovial cells but it is unusual to observe mitosis despite the attrition which is usually considered to occur in this membrane. None of the later electron microscopic studies have confirmed the presence of the large numbers of degenerating cells in the synovial membrane described by Lever and Ford (1958) and no studies of regeneration of the synovial surface using the electron microscope have yet been carried out. There is, however, little doubt that regeneration occurs rapidly after experimental synovectomy, but the character of the new cellular surface has not been elucidated.

The capillaries of the synovial membrane present no unusual features. All the most superficial vessels are devoid of either a muscular or elastic coat. These vessels approach within 10 to 20 μ of the joint surface. The findings agree with those of Lindström (1963). The two kinds of capillaries described by Suter and Majno (1964) with polarisation of the endothelial cell nuclei away from the joint cavity have not been confirmed in this material.

Summary

1. The synovium of the rabbit has been studied in glutaraldehyde fixed material.

2. A layer of amorphous material lies amongst and under the cells of the intimal layers. Its possible composition and function are discussed. Its role in the exchange from synovial fluid to blood or vice versa is considered.

3. The synovial cells cannot be distinguished as two morphological types, A and B, as suggested by several investigators. An assessment is made of present knowledge concerning the site of hyaluronate production.

4. Of the various organelles seen in the synovial cells the Golgi apparatus appears to occupy a prominent position. The endoplasmic reticulum is relatively scanty, mitochondria are few, and their cristae are scanty and small in most cases.

5. A flagellum is described on one cell in the subintimal layer.

We are indebted to the Arthritis and Rheumatism Council for the establishment of the Electron Microscope Unit, to Messrs. J. KING and G. MAXWELL for technical assistance, and to Mrs. E. M. SMITH for secretarial assistance.

References

BALL, J., J. A. CHAPMAN, and K. D. MUIRDEN: The uptake of iron in rabbit synovial tissue following intra-articular injection of iron dextran. J. Cell Biol. 22, 351—364 (1964).

BARLAND, P., A. B. NOVIKOFF, and D. HAMERMAN: Electron microscopy of the human synovial membrane. J. Cell Biol. 14, 207—220 (1962).

BARNETT, C. H., and A. J. PALFREY: Absorption into the rabbit articular cartilage. J. Anat. 99, 365—375 (1965).

BAUER, W., M. W. ROPES, and H. WAINE: The physiology of articular structures. Physiol. Rev. 20, 272—312 (1940).

CASTOR, C. W.: Production of mucopolysaccharides by synovial cells in a simplified tissue culture medium. Proc. Soc. exp. Biol. (N.Y.) 94, 51—56 (1957).

CHAPMAN, J. A., K. D. MUIRDEN, J. BALL, and P. A. HYDE: Synovial tissue and the uptake of iron following intra-articular injection. In: Proc. Fifth Internat. Congr. of Electron Microscopy, SS-12, S. S. BREASE Jr. (ed.). New York: Academic Press 1962.

COCHRANE, W., D. V. DAVIES, and A. J. PALFREY: Absorptive functions of the synovial membrane. Ann. rheum. Dis. 24, 2—15 (1965).

COULTER, W. H.: The characteristics of human synovial tissue as seen with the electron microscope. Arthr. and Rheum. 5, 70—30 (1962).

ENGEL, D.: The permeability of the synovial membrane. Quart. J. exp. Physiol. 30, 231—244 (1940).

— The influence of the sympathetic nervous system on capillary permeability. J. Physiol. (Lond.) 99, 161—181 (1941).

FRASER, J. R. E., and K. J. CATT: Human synovial cell culture. Use of a new method in a study of rheumatoid arthritis. Lancet II, 1437—1439 (1961).

GROSSFIELD, H., K. MEYER, and G. GODMAN: Differentiation of fibroblasts in tissue culture, as determined by mucopolysaccharide production. Proc. Soc. exp. Biol. (N.Y.) 88, 31—35 (1955).

HAMERMAN, D., and M. BLUM: Histologic studies on human synovial membrane. 2. Localisation of some oxidative enzymes in synovial membrane cells. Arthr. and Rheum. 2, 553—558 (1959).

—, and J. RUSKIN: Histologic studies on human synovial membrane. 1. Metachromatic staining and the effect of streptococcal hyaluronidase. Arthr. and Rheum. 2, 546—552 (1959).

HEDBERG, H., and V. MORITZ: Biosynthesis of hyaluronic acid in tissue cultures of human synovial membrane. Proc. Soc. exp. Biol. (N.Y.) 98, 80—88 (1958).

Hyde, P. A.: An electron microscope study of synovial membrane. M.Sc. Thesis, University of Manchester 1964.

Kling, D. H., M. G. Levine, and S. Wise: Mucopolysaccharides in tissue cultures of human and mammalian synovial membrane. Proc. Soc. exp. Biol. (N.Y.) **89**, 261—263 (1955).

Langer, E., u. F. Huth: Untersuchungen über den submikroskopischen Bau der Synovialmembran. Z. Zellforsch. **51**, 545—559 (1960).

Lever, J. D., and E. H. R. Ford: Histological, histochemical and electron microscopic observations on synovial membrane. Anat. Rec. **132**, 525—539 (1958).

Lindström, J.: Microvascular anatomy of synovial tissue. Acta rheum. scand., Suppl. 7, 1—82 (1963).

Palfrey, A. J., and D. V. Davies: The fine structure of chondrocytes. J. Anat. **100**, 213—226 (1966).

Richardson, K. C., L. Jarett, and E. H. Finke: Embedding in epoxy resins for ultrathin sectioning in electron microscopy. Stain Technol. **35**, 313—323 (1960).

Sabatini, D. D., K. G. Bensch, and R. J. Barrnett: Preservation of ultrastructure and enzymatic activity of aldehyde fixation. J. Histochem. Cytochem. **10**, 652—653 (1962).

— — — Cytochemistry and electron microscopy. The preservation of cellular ultrastructure and enzymatic activity by aldehyde fixation. J. Cell Biol. **17**, 19—58 (1963).

Sorokin, D.: Centrioles and the formation of rudimentary cilia by fibroblasts and smooth muscle cells. J. Cell Biol. **15**, 363—377 (1962).

Suter, E. R., and G. Majno: Ultrastructure of the joint capsule in the rat: Presence of two kinds of capillaries. Nature (Lond.) **202**, 920—921 (1964).

Thomas, D. Page, and J. T. Dingle: Studies on human synovial membrane *in vitro*. The metabolism of normal and rheumatoid synovia and the effect of hydrocortisone. Biochem. J. **68**, 231—238 (1958).

Vaubel, E.: Form and function of synovial cells in tissue culture. J. exp. Med. **58**, 63—83 (1933a).

— The production of mucin. J. exp. Med. **58**, 85—95 (1933b).

Wyllie, J. C., R. H. More, and M. D. Haust: The fine structure of normal guinea pig synovium. Lab. Invest. **13**, 1254—1263 (1964).

Yielding, K. L., G. M. Tomkins, and J. J. Bunim: Synthesis of hyaluronic acid by human synovial slices. Science **125**, 1300 (1957).

Biomechanics and Functional Adaption of Tendons and Joint Ligaments

A. Viidik

Introduction

In the present society of ever increasing mechanization the human body is subjected more and more to various mechanical stresses, which are resisted primarily by connective tissue in various forms. Interest has also been focused on connective tissue adaption to different functional states and to the changes induced by aging.

This paper deals with certain aspects of the dense regular form of soft connective tissue that constitutes tendons and ligaments. First, a survey is given of current knowledge of the morphology (down to the molecular level), the tensile strength, the elasto-plasticity characteristics, and the functional adaption and aging of these structures. Methods of studying different elasto-plasticity characteristics of ligaments are presented together with results from tests of the anterior cruciate ligament from the knee joint of rabbits. Some results from tests of the tensile strength characteristics of the Achilles tendon of rabbits and the anterior cruciate ligament in trained and untrained animals are presented together with methods suitable for these studies. Finally, the results are discussed in light of earlier investigations.

Survey of Literature

Morphology

The morphology of dense regular connective tissue such as occurs in tendons and joint ligaments is described in most textbooks of histology (e.g., BLOOM and FAWCETT, 1962; BUCHER, 1962) and has recently been reviewed from a functional point of view by HALL (1965). The tissue is composed of cells and intercellular substance, the latter consisting of fibers and ground substance. The fibers are mainly collagenous although some elastic and reticular ones also occur.

The cells in the mature tissue are fibrocytes. The fibrocytes, best studied in loose tissue, are usually described as having a nucleus that is spindle-shaped in profile and round in cross section, with the surrounding cytoplasm floating out in poorly defined processes which are difficult to stain histologically. The oval, centrally situated nucleus is described as having scattered chromatin and one or more large (BLOOM and FAWCETT, 1962; HALL, 1965) or small (BUCHER, 1962) nuclei. In tendons these cells are in parallel rows between the fiber bundles. Because the space there is quite restricted the cells often have a stellate appearance in cross sections but generally appear rectangular in profile. When seen in profile, the border between two cells is quite distinct but in cross sections the cytoplasmic processes extend between the fiber bundles as thin membranes that are difficult to differentiate from those coming from cells in other rows.

The collagen fiber is the most abundant type of fiber in the majority of tendons and ligaments. Young fibrocytes play a principal role in the formation of the collagen fibers. Unfortunately, in British and American literature, fibroblast is often used synonymously with fibrocyte. Recent authors support the extracellular aggregation hypothesis on collagen formation (ZELANDER, 1959; KARNER, 1960; INGELMARK, 1961; CASTOR and MUIRDEN, 1964; ZARZYCKI, 1964). The tropocollagen molecule or possibly free alpha chains are secreted by the fibrocytes and associate extra-cellularly (ROBERTSON, 1964). Some workers consider that the aggregation, in the form of extremely thin fibrils, occurs within the cells near the cell membrane. The primary building units of these fibers are the collagen molecules: three chains of amino acids in certain sequences which coil into lefthand helices and intertwine to form a righthanded superhelix (GROSS, 1961). Most investigators agree with this concept (RICH and CRICK, 1955; RAMACHANDRAN, 1956, 1962; HARKNESS, 1961; GROSS, 1961, 1964; GALLOP, 1964) although some advocate a single-helix model (GUSTAVSON, 1956; HUGGINS, 1962). The superhelix is the tropocollagen molecule from which various aggregation types of collagen can be formed: native collagen consisting of long chains with a periodicity of 700 Å, or the in vitro forms FLS (fibrous long spacing) and SLS (segment long spacing) with a major repeating periodicity of 2800 Å. The forms can, by laboratory procedures, be transformed from one to the other (GROSS, 1961). The sequence of amino acid in the collagen molecule is very orderly: two of the chains in the superhelix are of one variety, alpha-1, and the third of another, alpha-2 (GALLOP, 1964).

The amino acid compositions of the tropocollagen molecule in different mammals are amazingly similar although some slight differences occur in different tissues even from the same species (ROBERTSON, 1964). Proline and hydroxyproline occur frequently. Hydroxyproline is unique to collagen and together with proline, con-stitute about 25 per cent of the amino acid content of collagen. The collagen content of a sample can therefore be determined simply by a quantitative colorimetric measurement of its hydroxyproline content (STEGEMANN, 1958).

There also are some elastic fibers in dense connective tissue, mostly in the endotenon and "tenons" of higher orders and in elastic ligaments such as the ligamentum nuchae. This fiber is discussed in detail by HALL (1956).

The third type of fiber in connective tissue, the reticular fiber, is most common in loose connective tissue. The ground substance is less abundant in tendons and ligaments than in other connective tissue. It contains protein, mucopolysaccharides and glycoprotein and is a hydrophilic gel (CHRISMAN, 1964).

Collagenous tissue can be divided into "units" on different dimensional levels, by utilizing light and electron microscopy, but there is considerable confusion about the nomenclature and the actual dimensions of these "units." Some of these differences may perhaps be accounted for by the fact that collagen fibers from different species and different tissues are of varying dimensions (HARKNESS, 1961). Thus, the diameter of tendon fibrils is from 300—1300 Å in man (PAHLKE, 1954) and from 1600—4000 Å in the kangaroo (SCHMITT et al., 1942). How much these figures are influenced by different techniques in preparation of the specimens is difficult to estimate.

The following system of "units" seems most natural to the present author. The smallest "units" of collagenous tissue visible with the light microscope are the fibers which appear somewhat colorless, slightly refractive, and with a diameter ranging down to 0.2 μ. In polarized light microscopy the birefringence, an important feature, is displayed with hints towards the arrangement of more minute structures (BLOOM and FAWCETT, 1962). With the electron microscope the cross-striated fibrils are revealed. These fibrils have a diameter of about 200—600 Å and cross-striations with a periodicity of about 700 Å. By proper treatment protofibrils with a diameter of 15 Å, i.e., one molecule in width, can be precipitated from solutions of collagen.

The collagenous fibers are presumably as long as the tendon (BUCHER, 1962) and anastomose with each other at acute angles. The fibers are assembled into primary bundles that in large tendons may have a slightly helical form. The bundles are surrounded by a woven mesh of loose connective tissue, the *peritendineum internum* or endotenon, containing elastic fibers that tend to draw the bundle into a wavy formation when relaxed (RIGBY et al., 1959; WALLS, 1960). VERZÁR (1957) considered the structure to be helically arranged.

The whole tendon is surrounded by a sheath, the *peritendineum externum* or epitenon. Around the tendon is loose, fatty areolar tissue, the paratenon, in the meshes of which is mucinous fluid. The paratenon allows the tendon to glide freely against the surrounding tissue (BRAND, 1961; STANISAVLJEVIC and POOL, 1962). When the tendon passes over a structure that might harm it, a synovial sheath or a bursa is formed. NISBET (1960) considered this type of tissue to develop around the Achilles tendon of the rabbit, but the findings of VIIDIK (unpublished data) are more in agreement with those of LANG (1960), BRAND (1961) and STANISAVLJEVIC and POOL (1962), who considered the structure to be more loosely organized.

A ligament has a structure resembling that of a tendon although at times it is less orderly built (BLOOM and FAWCETT, 1962). Tendon fibers go directly into the bone and intermingle with the fibers of Sharpey (VIS, 1957; VIIDIK, 1964). This is also true for ligaments such as the cruciates in the knee joint (VIIDIK et al., 1965). In the transitional portion the cells undergo transition from fibrocytes to osteocytes and here they most resemble chondrocytes (VIIDIK, 1964; HALL, 1965; VIIDIK et al., 1965).

In the musculo-tendinous junction there is an intimate relationship between the muscle fibers and collagenous fibril bundles although there is no direct continuation. Electron microscopic studies have shown that the sarcolemma is intact at the junction and that tendon bundles are invaginated into the ends of muscle fibers in the many terminal indentations of the outer sarcolemmal layer. Thus, a considerable contact surface between the muscle fiber and the collagenous fibrils is achieved (GELBER et al., 1960; SCHWARZACHER, 1960).

Tensile Strength and Elasto-Plasticity Characteristics

Various investigators have ventured into this field, and a considerable amount of literature is available. The results are sometimes difficult to evaluate and contradictory at points.

Most authors agree that soft collagenous tissue, as in tendons and joint ligaments, is extensible and within certain limits elastic (WERTHEIM, 1847; ANNOVAZZI,

1928; GRATZ, 1931; STUCKE, 1950; DICK, 1951; SMITH, 1954; RIGBY et al., 1959; VIIDIK et al., 1965).

The most difficult parameter to measure has been the ultimate tensile strength. In order to determine this characteristic of a ligament or a tendon, one must be able to test it in an unobjectionable way with regard to the mounting in the testing machine. Also, if comparisons with preparations of different sizes are to be made, its cross-sectional area at the breaking point must be measured accurately. Tendons and ligaments are by no means homogenous structures and great attention must be given to the biological variations.

It must also be remembered that the tendon is a part of a biological system consisting of bone-tendon-muscle-tendon-bone and the ligament of a bone-ligament-bone complex and that they never function individually. MACMASTER (1933) showed on Achilles tendon complexes from rabbits that at least one-half of the tendon fibers had to be severed to make the tendon itself rupture when subjected to heavy loading. However, in other experiments failure in the system occurred at the Achilles tendon insertion (i.e., a tear-off fracture on calcaneus), the musculo-tendinous junction, the muscle belly, and the muscle origin. These findings have been confirmed by later experiments of other investigators (FINK and WYSS, 1942; STUCKE, 1951; DAVIDSSON, 1954, 1956).

Therefore, when the ultimate tensile strength of a tendon or a ligament complex is considered, factors other than the soft connective tissue must be taken into account The tensile strength of compact bone is less than its compressive strength (PEDER- SEN et al., 1949) and is probably less than of ligaments (BARNETT et al., 1961, p. 82)- Furthermore, the functional state of the different components must be considered

Table 1. *The ultimate tensile strength of tendons, in kiloponds per square millimeter, found by different investigators. All specimens are from unembalmed human tissue except those of* WALKER et al. (1964), *which are embalmed, and the second group of* WERT- HEIM, *which is from dogs. (When the values were given in pounds per square inch, the present author converted them to kiloponds per square millimeter)*

Author	Mean	Range
WERTHEIM (1847)	6.1	4.9—10.4
WERTHEIM (1847)	5.5	5.1—6.0
CRONKITE (1936)	—	6.1—12.7
STUCKE (1950)	4.7	— —
GUSTAVSON (1956)	—	10—12
WALKER et al. (1964)	—	7.5—15.0
WRIGHT and RENNELS (1964)	—	2.1—2.8

The results of different investigations on ultimate tensile strength of tendons are given in Table 1. Most authors agree that the average breaking load of a fresh tendon is about 4.5—6.0 kilopond[1] per square millimeter. The figures of BRAAMS (1961) are given as kiloponds per weight per length of tendon and the calculations to kp per mm^2 are performed by WALKER et al. (1964). The latter did not state their method and their figures are dubious. Their material consisted of tissues embalmed in a mixture containing, among other things, formaldehyde and phenol. It has been shown by several authors, e.g., STUCKE (1950), CURTIS (1963), and VIIDIK and LEWIN (1966), that chemical treatment of collagenous tissue alters its mechanical properties. Therefore, the results of WALKER et al. (1964) are difficult to evaluate.

---

[1] The force acting on the mass of one kilogram subjected to a gravity of 9.80665 meters/ sec^2 = one kilopond.

WRIGHT and RENNELS (1964) tested specimens of plantar fascia from amputated ischemic legs. They found the tensile strength varied from 2.1—2.8 kp per mm². All of their six specimens broke in the serrated jaws. Their results, judged by those of other investigators, must be influenced by their specimens and/or testing equipment errors.

It is generally considered that the stress-bearing structures on the molecular level are the collagen protein chains.

In their study of the action of hyaluronidase on rat-tail tendon fibers PARTINGTON and WOOD (1963) found that pure preparations of the enzyme had no significant effect on the load-extension curves. They concluded that chondroitin sulphate A and C and hyaluronic acid, the main components in the ground substance, had no part in the stability of the curve. However, no studies on the tensile strength were performed.

The behavior of collagen in tensile strength experiments is ultimately dependent on the number, types, and sites of the different inter- and intra-molecular bonds. It has been calculated (GUSTAVSON, 1956, p. 147) that breaking the chain at its weakest point, the —C—N— bond, would require 300 kp per mm² although the actual tensile strength is less than 15 kp per mm². Therefore, it is more probable that failure occurs in the transversal dimension where mostly cohesive forces exist.

It has been shown in studies of experimental lathyrism (van HEECHERAN and GROSS, unpublished data, cited by GROSS, 1964) that the loss of tensile strength of a tendon is a direct function of the chemical extractability of collagen.

That application of a load profoundly affects not only the macro- and microscopic structures but also structures on molecular level is indicated by changes in the wide-angle x-ray diagram when an extension of 10% of the tendon is performed (COWAN et al., 1953; KUHNKE, 1962).

The properties of the structures on the molecular level are influenced by cohesive and dispersive forces (LUNDGREN, 1949). The cohesive forces are the cohesion of the structure itself and, to some extent, the surrounding hydrostatic pressure. The dispersive forces are osmotic swelling, thermal force, and mechanical deformation. The profound influence of increased water content of collagenous tissue on its mechanical properties was shown by VIIDIK and LEWIN (1966).

The stress-strain curve or some approximation of it for tendons and ligaments has been described by several authors. Its significance and a special case, the load-elongation curve, have been discussed at length by VIIDIK et al. (1965).

Most authors agree that the stress-strain curve (Fig. 7) has at its beginning a toe part that is convex towards the load or stress axis (REUTERVALL, 1921; ÅKERBLOM, 1948; STUCKE, 1951; GUSTAVSON, 1956, p. 144; RIGBY et al., 1959; MORGAN, 1960). This agrees with the load-elongation curves, types A and B, of VIIDIK et al. (1965). WERTHEIM (1847) stated that the curve described a second-degree equation, although he did not mention the direction of the convexity. In contrast, the curves reproduced by WALKER et al. (1964) fail to display this feature. The toe part is considered to be caused by the wavy formation that the collagenous bundles display in a relaxed preparation (REUTERVALL, 1921; RIGBY et al., 1959). This formation is most easily seen in a polarizing microscope and vanishes when the stretching commences (RIGBY et al., 1959). The toe part is followed by the longest portion of the curve which is fairly linear. The structure may fail while the curve is in its linear part (as in the types B and C of Viidik et al., 1965) or it may level off towards the load/stress axis (type A). The rupture may be sudden, e.g., when the fibers fail simultaneously or when a preparation with bony attachments is tested and a tear-off fracture occurs; or the rupture may occur in successive steps, when the fiber bundles split at the weakest point in fairly rapid succession (Stucke, 1950; Viidik et al., 1965).

The elastic stiffness, i.e., the slope of the curve, is increased by drying (Wertheim, 1847; Walker et al., 1964) or vegetable and crome tanning (Gustavson, 1956, p. 144). Increased stiffening also occurs after brief treatment with formaldehyde (Curtis, 1963), but with prolonged treatment the elastic stiffness is less than that of native tissue (Curtis, 1963; Viidik and Lewin, 1966).

When a continuously increasing load is applied to the test specimen, dips sometimes appear in the load curve, i.e., slight drops of the load after which the curve more or less resumes its previous slope (Viidik et al., 1965). The dips have been interpreted as successive failures of the fibers stretched the most whereafter the load has been shifted to other strands of fibers until they in turn fail. This is to be expected from a morphological point of view since ligaments (and tendons as well) insert on areas and not on points. Therefore, no truly uniform tension can occur in any position of a joint (Barnett et al., 1961, p. 211).

The energy required to produce a failure in a test specimen can be calculated from the area below the load-elongation curve. This is of value when trying to assess differences between groups (Viidik et al., 1965).

A structure is said to be elastic if it returns to its original geometrical shape after the stress has been removed. If it doesn't return, it is plastic (or viscous) and the term elastic aftereffect means a gradual return to its original shape.

Ligaments from dogs were found to be elastic within certain limits by Annovazzi (1928). The limit for elasticity was lower in structures which also contained elastic fibers (interspinous and supraspinous ligaments in the cervico-dorsal region) rather than in purely collagenous ligaments from the knee joint. However, Gratz (1931), in his experiments on human fascia latae, found that the elasticity was about 90 per cent if the stress did not exceed about 1.4 kg per mm^2 (2000 psi) and less with greater stresses. This applied to fresh material and the results from "dead" tissue were less favorable. Stucke (1950) found a certain imperfection in the elasticity of human Achilles tendons. Smith (1954), on the other hand, stated that the anterior cruciate ligament from the knee joint of the rabbit was completely elastic with loads, of 5 minutes' duration, equal to the animal's body weight and with momentary submaximal loads. However, the ligament was plastic when submaximal loads were applied for a longer time.

Rigby et al. (1959), prior to their main experiments on rat tail tendon fiber bundles, performed a preliminary conditioning stretching which caused a slight permanent distension. They believed this resulted from "rupture of other components of the tendon (than the collagenous fibers) which take no subsequent part in the mechanical behavior." Thereafter, the tendon was perfectly elastic as long as the strain did not exceed 4 per cent, after which the plastic range commenced.

Wright and Rennels (1964) also noticed a permanent elongation of their specimens although they claimed elastic recovery from loads of 60—80 per cent of the failure load (i.e., from 1.8 kp per mm^2).

All published stress-strain data on collagenous tissue, displaying both loading and unloading curves on elasticity tests (GRATZ, 1931; STUCKE, 1950; RIGBY, 1964), have the toe part of the unloading curve closer to the strain axis and steeper than the corresponding loading curve. The area between the two curves represents the energy lost in such a cycle, but none of the papers deal with its possible significance or its variations.

Functional Adaption and Aging

Tendons and joint ligaments are among the most static tissues in the body with regard to their metabolism. Once collagen is mature, its turnover is very slow (NEUBERGER and SLACK, 1953; THOMPSON and BALLOU, 1956). The incorporation of amino acids into connective tissue is also comparatively slow and still slower in old animals (NEUBERGER et al., 1951; NEUBERGER and SLACK, 1953). Neutral salt soluble collagen, i.e., the youngest variety, has the greatest turnover (TAKASHIMA and TAMIYA, 1964) but is a living tissue that grows, ages, and responds to various stimuli.

Because the growth rate of intercellular substances exceeds that of the cells, the number of cells per unit of volume decreases rapidly after birth (INGELMARK, 1945, 1961). In man growth ceases at about 20 years of age. The internal and external peritendineum constitute 25 per cent of the tendon area in the newborn compared with only 12 per cent in the adult (INGELMARK, 1945). Tissue from older animals contains more collagen (GUSTAVSON, 1956, p. 30) but the water content decreases independently of the functional state of the tissue (INGELMARK, 1945). The degree of orientation of collagenous structures also tends to increase with age.

INGELMARK (1945) showed that when rabbits were subjected to training, the cross-sectional areas of the primary bundles in their Achilles tendons and muscles increased more than the number of cells. Training enlarged the Achilles tendons of young white mice but only the corresponding muscles of mature mice. This means that with increasing age enlargement of the muscle exceeds that of the tendon when subjected to strain (INGELMARK, 1948a). Electron microscopy showed the collagen fibrils were thicker in trained animals (INGELMARK, 1948b) but their dimensions varied more in older animals (VOGEL, 1964).

Little research has been done on the effect of immobilization on tendons and ligaments. The neutral salt soluble fraction of collagen is unchanged or reduced by immobilization (BROOKE and SLACK, 1959; PEACOCK, 1963) while the hydrothermal shrinkage characteristics are unaltered (PEACOCK, 1963; AKESON, 1963). Immobilization decreases the total hexosamine and water concentrations (AKESON, 1961), as well as the acid mucopolysaccharides (AKESON and LAVIOLETTE, 1964), but accentuates the metabolic turnover (LINDSAY et al., 1964). Probably the collagen in mature tissue is unaffected although no literature on its tensile strength characteristics is available.

KRATKY et al. (1962) in their x-ray diffraction studies on human tendinous tissue found no influence of immobilization or overstraining. However, their material is too meager to permit definite conclusions.

Connective tissue is also sensitive to different hormones. Thus HERRICK (1945) showed on castrated chickens that the tensile strength of muscles and skin was doubled when the birds were injected with testosterone. Estrogens exercise their primary influence on special areas and MULLER (1951) described the histology of the effect on loose connective tissue. Ground substance increases and there probably is also new formation of collagenous tissue (HALL, 1965). Relaxin and related substances influence mainly the connective tissue of the pubical symphysis (HISAW and ZARROW, 1950; STOREY, 1957).

The regeneration of tendinous tissue is rather slow (MASON and SHEARON, 1932; DAVIDSSON, 1956; BARNETT et al., 1961, p. 144). The healing begins from the sheath layers (LINDSAY and THOMSON, 1960), especially the paratenon (SCHNEE-WIND et al., 1964), and the collagenous fibers orient themselves in the direction of stress.

Aging has been studied with various techniques. It is generally considered that the neutral salt soluble fraction of collagen is the youngest and matures to the insoluble type via the acid soluble one (HALL, 1965, p. 19) although this concept is questioned (TSURUFUJI and OGATA, 1964). The tissues seem to lose their elasticity (BARNETT et al., 1961, p. 138), an effect that can also be produced by exposing young tendon fibers to diluted formaldehyde (ELDEN, 1965).

Molecular changes seem to be profound as indicated by x-ray diffraction and biophysical studies (KRATKY et al., 1962). Thermal shrinkage of collagen (VERZÁR, 1956), shrinkage induced by concentrated solutions of urea (ELDEN and BOUCEK, 1962; ELDEN, 1964), and potassium iodide (BANGA et al., 1956a, 1956b) increase considerably with aging. This is probably caused by increased cross-linkage between adjacent collagen molecules (VERZÁR, 1963; SINEX, 1964).

Collagen seems to age more rapidly in wild animals than in their domesticated counterparts (CHVAPIL and ROTH, 1964).

Present Investigations
Principal Remarks

The mechanical properties of biological specimens with collagen as one of the main components have been studied by numerous authors, biologists as well as physicists. The methods used have varied widely depending on the main interests of the investigators. Furthermore, the methods have not always been without objections from both of the above-mentioned different points of view.

This study is based upon observations of well-defined scientific phenomena. There are a large number of such phenomena in the field of mechanical properties of materials, e.g., elasticity, plasticity, creep, fatigue, and others, but a complete investigation of all of them was not made in the present study. The most interesting of them, from both biological and technical points of view, were selected for study here.

Complete functional units were used by numerous earlier investigators (e.g., MacMASTER, 1933; FINK and WYSS, 1942; STUCKE, 1951; SMITH, 1954; DAVIDSSON, 1954, 1956; VIIDIK et al., 1965). The other alternatives have been to use isolated and well-defined bundles of collagen fibers (e.g., RIGBY et al., 1959; CURTIS, 1963) or whole isolated tendons or ligaments (e.g., STUCKE, 1950; WALKER et al., 1964).

Instead of studying isolated specimens of relatively homogenous tissue, whole functional units, bone-ligament-bone and bone-tendon-muscle-tendon-bone preparations, were investigated here.

As a criterion for static behavior the load-elongation diagram of the femur-anterior cruciate ligament-tibia preparation was used, the ligament being the only structure exhibiting notable elongation during loading. The load-elongation diagram gives information on the elastic stiffness (tan α), the failure load, the corresponding elongation, the energy required for rupture, and so on. The ultimate conditions, however, are never reached in a living and normally functioning unit but are of interest when studying injuries.

As a criterion for dynamic behavior a fixed cycle of rapid loading and unloading, with a safely submaximal load, was chosen. The length was kept constant during loading and adjusted at certain time intervals to insure a fixed load. Although this was not a very true simulation of the real functional behavior of the tested unit, which may vary widely, the method gave information of damping and creep in the specimen, the latter being considered as a measure of the capacity of the biological material to recover. Because the number of repeated loadings was small, no more than four per specimen, the test procedure gave no information of the real fatigue conditions which were of minor interest.

Elasto-Plasticity of Rabbit Ligaments
Theoretical Background

Before describing the apparatus and techniques used in the study, the observed phenomena will be defined and, in some cases, elucidated by simple mechanical analogies.

The conception "dashpot," meaning a viscous damper, is used when the resistance to motion is proportional to the velocity of the motion.

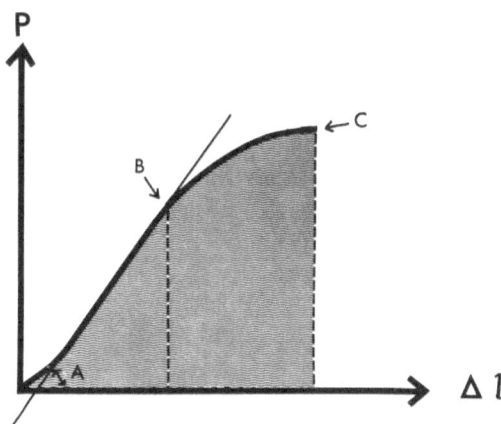

Fig. 1. A schematic load (*P*)-elongation (*Δ l*) diagram. The tangent for angle *A* gives the elastic stiffness within the elastic range. The expected beginning of the plastic range is indicated by *B* and the point of rupture by *C*. The shaded area represents the energy required to achieve failure

Stress-Strain Relation is basic for calculating the relation between deformation and load as well as the failure characteristics of a body of any shape.

Because the universal characteristics, stress (load per unit area) and strain (elongation per unit length), are most often difficult to measure accurately, the load-elongation relation is a good replacement for specimens of similar configuration and more so when congruent (cf. discussion by VIIDIK et al., 1965).

In Fig. 1 some interesting physical quantities are pointed out on a schematic load-elongation curve.

Elasticity Limit is the maximal stress-strain value at which the body after unloading has no residual deformation.

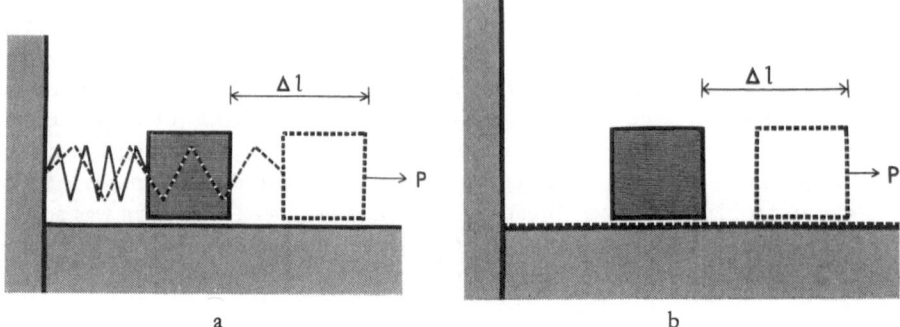

Fig. 2a and b. Mechanical analogies. a Illustrates perfect elasticity. The square on a non-frictional ground representing the specimen is attached to the wall via a spring. This gives a displacement of the specimen from the equilibrium that is proportional to the force applied. When the force is removed the specimen returns completely to the original equilibrium position. b Illustrates perfect plasticity. There is dry friction between the square and the ground which gives constant resistance against motion independent of direction. When a force has displaced the specimen, cessation of the force doesn't alter the displaced position of the specimen

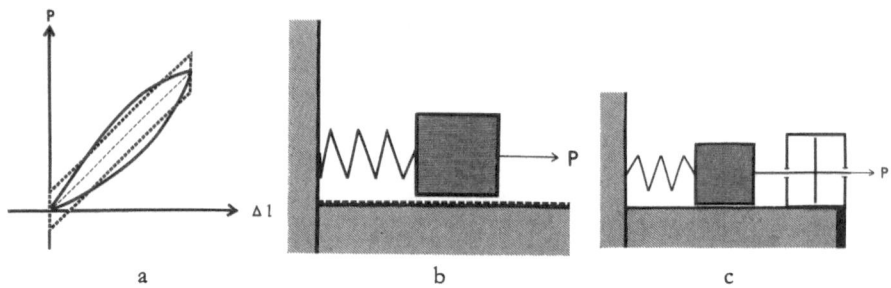

Fig. 3a—c. Load-elongation diagram and mechanical models for internal damping. See text for explanation. a The dotted curves represent a cycle of loading and unloading of the model in Fig. 3b, and the solid curves designate the same in Fig. 3c. The upper parts of the cycles are for loading and the lower ones for unloading. b A model of internal damping as the result of dry friction. The force has to overcome the elastic resistance from the spring and the dry friction between specimen and ground before any displacement can occur. In unloading, the frictional force alters direction and is thus antagonistic to the elastic force. c A model of internal damping as the result of viscous friction. This is visualized by a dashpot coupled between the acting force and the specimen that is perfectly elastic. The dashpot acts as a disturbing and counteracting force proportional to the velocity of the system. In the beginning and in the completion of loading or unloading movement the influence of the dashpot is zero because the displacement velocity is usually very small. See text for definition of dashpot

Perfect (Linear) Elasticity indicates a completely reversible proportionality between stress and strain (Fig. 2a). The slope of the linear part of the stress-strain or load-elongation curve (Fig. 1) gives the constant of proportionality, i.e., the modulus of elasticity.

Perfect Plasticity means that an irreversible constant stress is applicable independent of the strain (Fig. 2b). True plastic materials have irreversible stress-strain relations where the stress does not need to be constant.

Internal Damping or Hysteresis refers to a consumption of energy during a loading-unloading cycle. Hysteresis is the visible result of internal friction, which can be dry, viscous or of some other character. The dry friction (Fig. 3b) gives a load-elongation relationship shown by the dotted-line cycle in Fig. 3a. The viscous type, symbolized by the dashpot in Fig. 3c, gives the solid-line cycle in Fig. 3a. The areas within the dotted and the solid line curves, respectively, in Fig. 3a represent the amount of energy lost in friction work.

A dashpot is a viscous damper, in which the resistance to movement is proportional to the velocity. Thus, if a force is applied to its piston, acceleration to a constant velocity occurs exponentially.

Creep includes different types of elastic and plastic aftereffects. The simplest mechanical analogy of the creep phenomenon is given in Fig. 4. The analogy consists of a spring and a dashpot in parallel. (The model is, except for the dashpot, nonfrictional.)

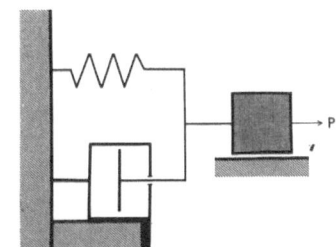

Fig. 4. A mechanical analogy for the creep phenomenon. See text. The specimen is placed on a nonfrictional ground and attached to the wall via a spring and a dashpot (cf. Fig. 3c) in parallel. The initial resistance to a force that is rapidly applied is caused mainly by the dashpot. After the force has acted for some time the spring is gradually elongated until the spring force is of the same magnitude as the load. Then the dashpot is unloaded and the velocity is zero. If on the other hand a rapid elongation is made, it creates a counterforce mainly in the dashpot. This ceases with time if the elongation persists and the remaining counterforce originates then from the spring

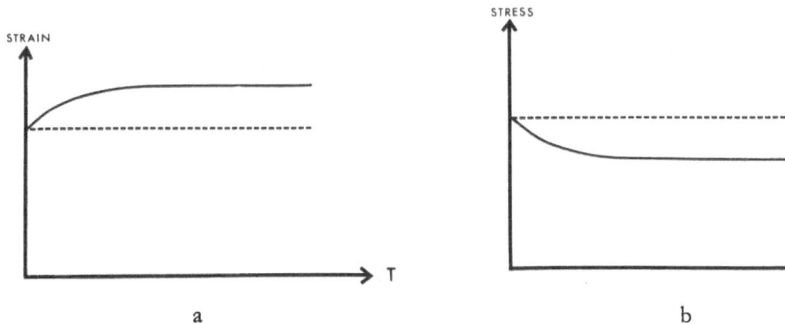

Fig. 5a and b. Creep phenomena visualized graphically. Cf. Fig. 4. a Shows the strain (elongation per original length)-time relationship when the stress is constant. b Shows the stress (load per unit area)-time relationship when the strain is constant

If a specimen with these combined properties is brought momentarily to a certain stress-strain level on the stress-strain curve and then exposed to either constant stress or constant strain, its time-dependent behavior is as in Fig. 5a and b, respectively. The curves behave like exponential functions.

Material and Methods

In this part of the study 18 knee-joint preparations from mature rabbits were used. The animals were sacrificed by heavy occipital blows and the tests performed within two hours after death.

Test specimens consisting of the tibia-anterior cruciate ligament-femur preparations were fastened into contour-shaped clamps as described by Viidik et al. (1965). The tests were performed according to the schedule given in Table 2.

Table. 2. *The schedule for the loading-unloading cycle experiments in the different groups. The load adjustments were to achieve the initial load*

| | Testing Group Number | | | | |
	A	B	C	D	E
Initial load up to	10 kp	10 kp	10 kp	10 kp	3 kp
1st load adjustment after	2 min	1 min	2 min	2 min	2 min
2nd load adjustment after	2 min	1 min	2 min	2 min	2 min
Then unloaded after	2 min	1 min	2 min	2 min	2 min
The loading-unloading cycle repeated after	10 min	5 min	60 min	240 min	10 min
No of cycles	3	4	2	2	3

A load of ten kiloponds was selected as a considerable load and yet safely below the tensile strength of the ligament which in such preparations is about twenty-seven kiloponds with a variation downwards to twenty kiloponds. A load of three kiloponds was chosen to correspond to the load of the body weight used by Smith (1954).

During the tests the ligament was kept moist by wrapping it in saline-moistened gauze and checking its condition from time to time. Preserving the moisture of ligamentous tissue by immersing it into fluid was rejected by Viidik and Lewin (1966).

Because the testing program demanded accurate adjustments a test rig, adjustable within very small limits, was designed to allow the specimen to be exposed to elongations. Thus, the elongation was considered to be the independent variable and the corresponding load the dependent variable.

The device for producing and measuring the elongation was designed to eliminate as much backlash in the system as possible. The mobile end of the clamps was connected via a doublesided axial and radial bearing to a shaft one end of which was threaded. The pitch of the thread was 1.5 mm and the threaded shaft operated inside a fixed nut.

In order to make the screw arrangement backlash-free, the nut was sliced into three radial sections, each of which was pressed against the thread of the shaft by an external nut.

The specimen was elongated by manual rotation of a wheel on the shaft, one revolution of the shaft elongating the specimen 1.5 mm. The periphery of the wheel was divided into fifteen parts, one part corresponding to 0.1 mm. Still better accuracy was attained with the aid of four subdivisions in each part (Fig. 6).

The motion of the mobile part of the system was recorded with a displacement transducer of differential transformer type (Bofors RLK-1-S). The transducer was fixed to the bottom plate of the testing device and its moving part, the iron core, fixed to the moving part via a micrometer screw used to calibrate the transducer and to zero the reading at the beginning of a test series.

Fig. 6. The apparatus for the elasto-plasticity tests. In the center of the transparent plastic box are the clamps which hold the femur-anterior cruciate ligament-tibia preparation and to their right is the force transducer. Inserted into the left side of the box is the plunger going to the back-lash free screw arrangement (to elongate the system being treated), with the displacement transducer attached to its left side. A measuring bridge and part of the recorder are seen in the upper-left portion of the photo

The displacement of the transducer was recorded continuously by a Philips PT 1200 direct reading measuring bridge. The corresponding load was measured at the fixed end of the system by a Bofors KRK-1 force transducer (max. load 20 kp) coupled to another direct reading measuring bridge. The corresponding load and elongation values were recorded simultaneously and continuously in two ways: (1) by a Siemens Oscillomink ink jet recorder on separate channels, i.e., the load and elongation on the ordinates and time on the abscissas; (2) by a Tectronix 502 oscilloscope X-Y-coupled with the load on the ordinate and the elongation on the abscissa, the tracing being photographed with a Rolleicord Camera adapted to oscilloscope photography (Philips PM 9300).

Before each test series mechanical calibrations were carried out by means of weights hanging down from the force transducer and displacement by means of the micrometer screw previously mentioned. Calibrations before and after a series of experiments displayed very small variations. Electronic calibration of the system was carried out by built-in calibration impulse units in the bridges and was found to remain constant.

Results

The general shape of the load-elongation curve of the loading was similar in all tests. It began with the earlier mentioned toe part, after which a fairly linear portion ensued. In none of the cases did the curve tend to level off at the top. The corresponding curve for unloading was similar although always steeper and with the toe part more accentuated. The second loading curve exhibited a longer and more accentuated toe part than the corresponding one from the first loading.

In the first loading of an anterior cruciate ligament preparation to ten kiloponds the tangent of the slope of the linear portion of the load-elongation curve (tan α) was 1.40 (kp per 0.1 mm) with a standard error of ±0.05. The corresponding figure in the tensile strength study by Viidik et al. (1965) was 1.53±0.08.

With repeated tests on the same specimen the tan α value tended to increase although no corresponding change was observed in the tan α values on the unloading sequence (Table 3). The difference between the tan α values on the loading and unloading was a constant feature: the unloading value was always higher. The difference was greatest in the first testing of a specimen but considerably smaller and rather constant in the following testings (Table 3). A typical example of the curves for testing the first and second times is seen in Fig. 7.

Table 3. *The slope of the load-elongation curve (tan α) for loading and unloading in elasto-plasticity test group B. The mean values and standard errors are given. The fourth column gives the mean values and standard errors for the differences between the pairs of loading and unloading curves.* * *denotes significant differences from cycle n:o 1*

Cycle n:o	Loading	Unloading	Difference
1	1.40±0.05	1.71±0.08	0.37±0.07
2	1.53±0.07	1.68±0.09	0.14±0.04 *
3	1.58±0.08 *	1.78±0.09	0.19±0.05
4	1.59±0.08 *	1.78±0.13	0.18±0.08
2+3+4	1.57±0.04 *	1.74±0.06	0.17±0.03 *

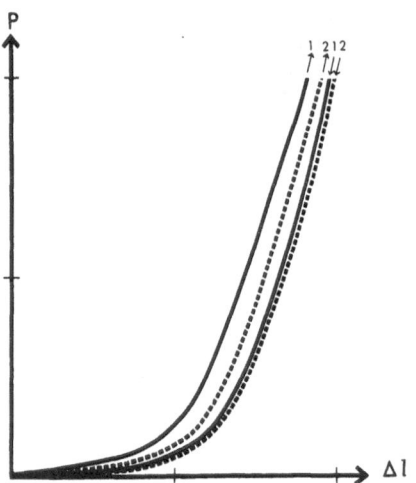

Fig. 7. Typical load-elongation curves from first (———) and second (— — —) tests of elasticity on an anterior cruciate ligament preparation. ↑ indicates loading and ↓ unloading

If the elongation required to reach an initial load of ten kp is set at 100 per cent, the second and third loadings give the values of 107 and 109 per cent, respectively. In the corresponding tests with loads of only three kp the elongation values of 100, 109, and 109 per cent were found.

With constant elongation the load diminished with time, most rapidly in the beginning and the curve (Fig. 5b and 8a) resembled an exponential function. The creep thus displayed was greatest in the first test of a specimen and decreased considerably in the second and all following tests (Fig. 8a). There was no difference in the creep of the second test if the resting period was ten minutes or four hours (Fig. 8b). The relative creep, in per cent of the initial load, did not differ if the load was ten or three kp (Fig. 8c).

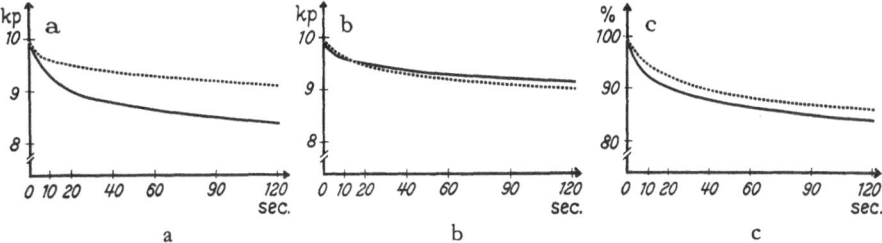

a   b   c

Fig. 8a—c. Tests on the creep phenomenon. a Gives the mean values of the creep during the first 120 seconds in the first (———) and second (———) test cycles of group A. b Gives the mean values of the creep during the first 120 seconds in the second test cycles of group A, i.e., after 10 minutes rest (———) and group D, i.e., after 240 minutes rest (———). c Gives the mean values of the creep during the first 120 seconds in the first test cycles in group A, i.e., 10 kp initial load (———) and group E, i.e., 3 kp initial load (———)

The Effect of Training on the Tensile Strength Characteristics of Rabbit Tendons and Ligaments

Theoretical Background

In order to be able to evaluate as many parameters as possible when performing tensile strength tests on Achilles tendons from rabbits, the author developed a methodology to test the whole complex: calcaneus-tendo Achillis-musculus gastrocnemius-femur.

For the applicability of the load-elongation diagram and the different parameters, see VIIDIK et al. (1965). The principle of electronic registration of load and elongation used in the present study is basically the same as theirs.

Material and Methods

Four rabbits were trained in a running machine every working day for eight months. They were growing when the training started and adult at the time of the experiments. During the training period they ran approximately 100 kilometers. Four rabbits from the same breed were kept as controls in cages during the training period. Both legs from each rabbit were used for the eight tests in each group.

In the preliminary experiments the animals were sacrificed in different ways. When agonal struggle of the hind limbs had occurred, the muscle belly often ruptured with a considerably lower load than when avulsion of the tendon occurred. The same was true when several hours lapsed between sacrificing and testing. The sacrificial method of choice was asphyxia caused by a tubocurarine injection (1 mg per kg body weight intravenously) after anesthesia induced with allypropymalum (0.5 ml 10 per cent solution per kg body weight intravenously).

The calcaneus was fitted in a contour-shaped clamp with a rounded outlet for the tendon and fastened to the immobile end of the materials testing machine. The femur was likewise fastened in a contour-shaped clamp, in the construction of which care was taken to avoid sharp edges. The clamp was fastened to the plunger, at the other end of the machine, worked by a hydraulic system. On the transition point between the tendon and muscle a steel ring was fastened (Fig. 9).

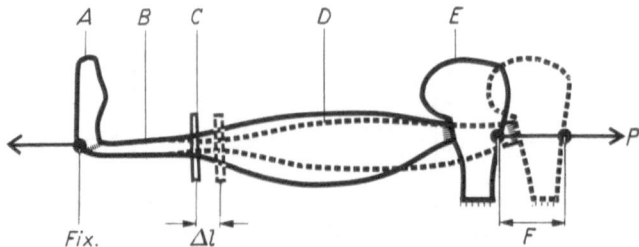

Fig. 9. A sketch of a tendon complex before (———) and during (– – –) tensile strength testing indicating: *A* the calcaneus; *B* the Achilles tendon; *C* the ring indicating the musculo-tendinous junction and the elongation (*Δl*) of the tendon when testing; *D* the gastrocnemius muscle; *E* the femur and *F* the elongation of the whole preparation

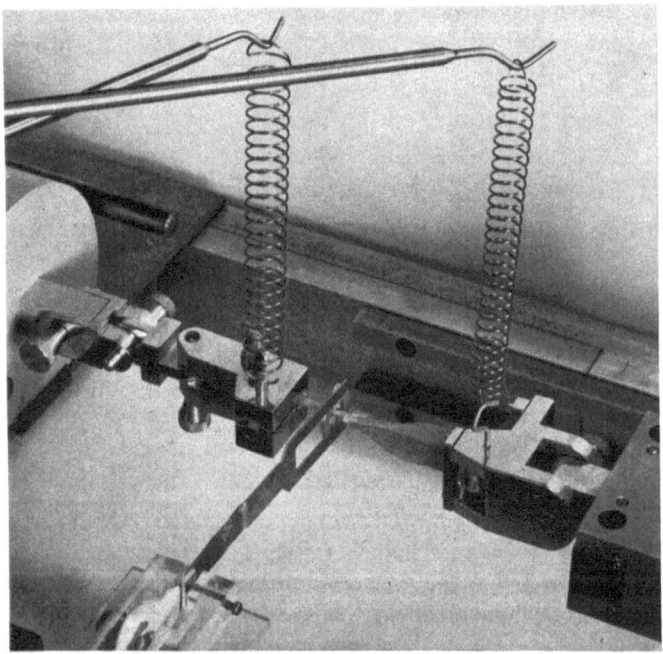

Fig. 10 shows an Achilles tendon preparation mounted for tensile strength testing. From the left: the tensile load pick-up, the calcaneus mounted in clamp, the Achilles tendon, the ring indicating the musculo-tendinous junction with the pliable U-shaped steel plate with the elongation reading strain gauges, the muscle, the femur in clamp and the hydraulic plunger

When loading the system the tendon and the muscle elongated and, therefore, the ring moved away from the immobile end of the machine. This movement indicates the elongation of the tendon. The U-shaped end of a pliable steel plate was placed against the ring and the displacement was registered by means of two Philips PR 9810 strain gauges cemented on the plate and connected in a half bridge circuit to a direct reading measuring bridge. The load was registered with a Philips PR 9226/02 (max. load 200 kp) tensile load pick-up connected to another bridge. The load was applied to the system until failure occurred (Fig. 10). The registrations and calibrations were carried out as described previously.

The preliminary experiments were recorded by cinematography (with a Paillard H16 on 16 mm Kodachrome, 64 pictures/sec). The analysis of the pictures, both sequentially and individually, revealed no slipping of the clamps or of the steel ring. No irregularities in the deflection of the steel plate by the steel ring at the muscle-tendon junction were found.

The tensile strength tests on the anterior cruciate ligaments of the knee joints were performed with the same equipment. The femur and tibia were fastened in contour-shaped clamps after all structures, except the ligament to be tested, were severed. The elongation was measured by strain gauges on a steel plate deflected by a steel pin at the moving end of the system.

Some of these latter experiments were recorded with a Kodak High Speed Camera at 1000 frames/sec. Analysis of these reels showed that during the loading the middle of the ligament became thinner and, as an effect of the elongation, the insertions seemed to broaden. The tear-off fracture from the tibia occurred very suddenly, in the course of one or two frames. In a few instances a partial tear in the ligament, corresponding to a dip in the load curve, was noticed prior to failure of the femur-ligament-tibia preparation that occurred at some other point.

The following parameters were studied: (1) the gross shape of the load-elongation curve, (2) its slope (tan α), (3) its failure energy, as represented by the area beneath the load-elongation curve, (4) failure load, (5) elongation at failure, (6) dips in the load curve, and (7) failure site on the bone-ligament-bone and bone-tendon-muscle-bone preparations.

The dips in the load curve were studied only in the ligament tests, because a corresponding analysis of the tendon tests could not differentiate between partial ruptures in the muscle belly and in the tendon.

Results

Cursory inspection of the oscillograms showed that all of the curves from the tensile strength tests on Achilles tendons exhibited a toe part in the beginning followed by a fairly linear portion which finally leveled off towards the load axis before failure occurred. The toe part seemed more accentuated in the control group and the bending off to the linear portion sharper in the trained group. The corresponding curves from tests on ligament preparations showed the same features but less sharply.

The slope of the linear portion of the load-elongation curve (tan α) was steeper for the trained than for the untrained tendons but no difference was discernable in the ligament tests (Table 4).

The failure energy, both in the ligament and tendon tests, was higher in the trained than in the control group, the difference between the two groups being greater in the ligaments (Table 4). The same was true for the failure load. The elongation at failure load was the same, however, in the trained and control groups both in the tendon and ligament tests. A curious feature was that in neither the

Table 4. *Mean values and standard errors for the different tensile strength parameters in trained and untrained rabbits*

Parameter	Tendons		Ligaments	
	Trained	Untrained	Trained	Untrained
tan α	1.72±0.42	1.16±0.20	2.26±0.19	2.16±0.19
Failure energy, kpmm	62.1±9.0	58.5±6.8	45.9±12.1	30.9±3.4
Failure load, kp	41.9±4.2	38.8±3.3	34.6±3.2	29.9±2.4
Elongation at failure load, mm	2.55±0.31	2.59±0.20	1.84±0.28	1.34±0.14
Maximal linear load, kp	25.9±3.8	25.2±4.1	25.1±2.1	25.1±2.2

ligament nor the tendon tests was there any difference between the trained and control groups in the length of the linear portion of the load-elongation curves.

In the tendon tests all preparations failed with a tear-off fracture from the calcaneus except one in the trained group which ruptured first at the insertion of the tendon from the lateral head of the gastrocnemius muscle whereafter the rest of the preparation failed in the muscle belly.

Table 5. *The frequency of dips in the load-time curves and their magnitudes in the ligament preparation tensile strength tests*

Dips		Trained	Untrained
Total n:o		11	13
Magnitude	Mean	1.4	1.6
in kp	Range 0.5—3		0.5—5

All the ligament preparations failed as tear-off fractures except two in the control group, which failed in the ligament. The frequency of the dips showed no significant differences between the trained and the control groups (Table 5).

The body weight in the trained group was 3.2±0.2 kg and in the untrained group 3.1±0.2 kg.

Discussion

In the literature the accounts of the elasticity and/or plasticity of collagen are somewhat contradictory. In the present investigations the load-elongation curve was similar to those found by earlier investigators (e.g., Stucke, 1950; Rigby et al., 1959). The most applicable mechanical analogy was one displaying the creep phenomenon (Fig. 4) with an element of plasticity (Fig. 2b) which was most evident in the first loading as compared with the following ones.

There was a clear difference (Table 4) between the slopes of the load-elongation curves in loading and unloading. The first cycle was distinguished from the following ones in that a significantly greater difference occurred. In the second and following cycles a somewhat extended elongation was required to achieve the initial maximal load that occurred in the first cycle. Calculated in percentage it made no difference if the load was three or ten kiloponds. This agrees with the results of Gratz (1931), Stucke (1950), and Rigby et al. (1959), but is contradictory to those of Smith (1954).

The underlying cause for the imperfection in the elasticity of the ligament is not yet clear. Rigby et al. (1959) suggested that it was dependent upon rupture of non-collagenous components in the structure although this has not yet been demonstrated by histological investigations. Drying of soft collagenous tissue increases its elastic stiffness and could possibly be a cause of elastic imperfection. This is not the case, because RIGBY et al. (1959) performed their tests with the specimens immersed in saline and in the present studies the ligaments were wrapped in saline-moistened gauze. Furthermore, if the increased elastic stiffness was caused by drying of the specimen, the unloading part of the curve would also have been affected, which was not the case.

The creep phenomenon was also studied and found to be considerably greater in the first testing cycle than in the following ones. The resting period between the first and second loading did not influence the creep during the second cycle. The creep was also the same, calculated in per cent of the initial load, regardless of whether the load was ten or three kiloponds.

This is interesting from the functional point of view and has not been described by earlier investigators. It supports the concept that something occurs during the first testing cycle that profoundly affects the collagenous structures. What this is cannot yet be stated; neither is it clear of what magnitude the forces are that operate within the structures during physiological functioning. The phenomena described here might be responsible for some varieties of ligament strains.

The author's experiments are not concerned with the ability of these structures to recover over a longer period of time because it was not feasible with the types of tests made in the studies.

Some tests comparing the tensile strength of rabbit tendons and ligaments in trained and untrained animals were performed. The morphological response of tendons to increased functional demands has been previously clarified (INGEL-MARK, 1945, 1948a, 1948b, 1961). Whether this morphological response is accompanied by an equivalent improvement in the biomechanical properties of the structures is uncertain. This study indicates that the insertion of the Achilles tendon on the calcaneus is the weakest point, although somewhat stronger, after systematic training. Moreover, the elastic stiffness tends to become higher, hinting at some changes in the weight-bearing properties of the tendon.

In the bone-ligament-bone system the attachments of the ligament to the bone are the weakest point although they tend to be stronger in trained animals. The elastic stiffness displays no great change between the trained and untrained animals.

The figures obtained on the effect of training are not sufficiently numerous to permit definite conclusions. A more complete study is in progress, the intention of which is to elucidate the biomechanical response not only of whole systems, such as bone-tendon-muscle-bone and bone-ligament-bone, but also that of single components such as isolated tendons.

Acknowledgement

I am greatly indebted to Lecturer M. MÄGI, Ph. D., Chalmers University of Technology, for his invaluable aid in the design of some parts of the apparatus and in the preparation of the discussion on mechanics.

References

ÅKERBLOM, B.: Standing and sitting posture — with special reference to the construction of chairs, 1. ed. Stockholm: AB Nordiska Bokhandeln 1948.

AKESON, W. H.: An experimental study of joint stiffness. J. Bone Jt Surg. A 43, 1022—1034 (1961).

— Relationship between the aging phenomena in connective tissue and the connective tissue response to immobility: a thermodynamic approach. Surg. Forum 14, 438—440 (1963).

—, and D. F. LaVIOLETTE: The connective tissue response to immobility. Total mucopolysaccharide changes in dog tendons. J. surg. Res. 4, 523—528 (1964).

ANNOVAZZI, G.: Osservazioni sulla elasticità dei legamenti. Arch. Sci. biol. (Napoli) 11, 467—501 (1928).

BANGA, I., J. BALÓ, and D. SZABÓ: Submicroscopic structure of collagen fibres; their contraction and relaxation. Acta morph. Acad. Sci. hung. 6, 391—403 (1956a).

— — — The structure, aging and rejuvenation of collagen fibres. Experientia (Basel), Suppl. 4, 28—31 (1956b).

BARNETT, C. H., D. V. DAVIES, and M. A. MacCONAILL: Synovial joints. Their structure and mechanics, 1. ed. London: Longmans 1961.

BLOOM, W., and D. W. FAWCETT: A textbook of histology, 8. ed. Philadelphia: W. B. Saunders Co. 1962.

BRAAMS, R.: The effect of electron radiation on the tensile strength of tendon. I. Int. J. Radiat. Biol. 4, 27—31 (1961).

BRAND, P. W.: Tendon grafting. Illustrated by a new operation for intrinsic paralysis of the fingers. J. Bone Jt Surg. B 43, 444—453 (1961).

BROOKE, J. W., and H. G. B. SLACK: Metabolism of connective tissue in limb atrophy in the rabbit. Ann. rheum. Dis. 18, 129—136 (1959).

BUCHER, O.: Histologie und mikroskopische Anatomie des Menschen, 3. ed. Bern: Hans Huber 1962.

CASTOR, C. W., and K. D. MUIRDEN: Collagen formation in monolayer cultures of human fibroblasts. Lab. Invest. 13, 560—574 (1964).

CHRISMAN, O. D.: The ground substance of connective tissue. Clin. orthop. 36, 184—193 (1964).

CHVAPIL, M., and Z. ROTH: Connective tissue changes in wild and domesticated rats. J. Geront. 19, 414—418 (1964).

COWAN, P. M., A. NORTH, and J. T. RANDALL (1953): Cit. by GUSTAVSON (1956).

CRONKITE, A. E.: The tensile strength of human tendons. Anat. Rec. 64, 173—186 (1936).

CURTIS, D. H.: The effect of chemical crosslinking agents on the mechanical properties of rat-tail tendon, 1. ed. Ann Arbor: University Microfilms Inc. 1963.

DAVIDSSON, L.: Tensile strength, rupture and regeneration of tendons. Ann. Chir. Gynaec. Fenn. 43, 61—66 (1954).

— Über die subkutanen Sehnenrupturen und die Regeneration der Sehne. Eine experimentelle, klinische und pathologisch-anatomische Untersuchung. Ann. Chir. Gynaec. Fenn., Suppl. 6, 1—113 (1956).

DICK, J. C.: Tension and resistance to stretching of human skin and other membranes. J. Physiol. (Lond.) 112, 102—113 (1951).

ELDEN, H. R.: Aging of rat tail tendons. J. Geront. 19, 173—178 (1964).

— Elasticity of aging tendons. Collagen Curr. 5, 484 (1965).

—, and R. J. BOUCEK: Investigations of the aging process by physical-chemical means — summary. In: Biological aspects of aging, p. 334—342 [N. W. SHOCK (ed.)]. New York: Columbia University Press 1962.

FINK, R., u. O. A. M. WYSS: Experimentelle Untersuchungen über Rupturen am Knochen-Sehnen-Muskel-System beim Frosch. Mschr. Unfallheilk. 49, 379—389 (1942).

GALLOP, P. M.: Concerning some special structural features of the collagen molecule. In: Connective tissue — intercellular macromolecules, p. 79—92. Boston: Little, Brown & Co. 1964.

GELBER, D., D. H. MOORE, and H. RUSKA: Observations of the myotendon junction in mammalian skeletal muscle. Z. Zellforsch. 52, 396—400 (1960).

GRATZ, C. M.: Tensile strength and elasticity tests on human fascia lata. J. Bone Jt Surg. **13**, 334—340 (1931).

GROSS, J.: Collagen. Sci. Amer. **204** (5), 120—130 (1961).

— Organization and disorganization of collagen. In: Connective tissue — intercellular macromolecules, p. 63—77. Boston: Little, Brown & Co. 1964.

GUSTAVSON, K. H.: The chemistry and reactivity of collagen, 1. ed. New York: Academic Press 1956.

HALL, D. A.: Chemical studies on the relationship between elastin and collagen. Experientia (Basel), Suppl. **4**, 19—27 (1956).

HALL, M. C.: The locomotor system. Functional histology, 1. ed. Springfield (Ill.): Ch. C. Thomas 1965.

HARKNESS, R. D.: Biological functions of collagen. Biol. Rev. **36**, 399—463 (1961).

HERRICK, E. H.: Tensile strength of tissues as influenced by male sex hormones. Anat. Rec. **93**, 145—149 (1945).

HISAW, F. L., and M. X. ZARROW: The physiology of relaxin. Vitam. and Horm. **8**, 151—178 (1950).

HUGGINS, M. L.: Some developments concerned with the structure of collagen. In: Collagen, p. 79—80 [N. RAMANATHAN (ed.)]. New York: Interscience Publ. 1962.

INGELMARK, B. E.: Über den Bau der Sehnen während verschiedener Altersperioden und unter verschiedenen funktionellen Bedingungen. Upsala Läk.-Fören. Förh., N.F. **50**, 357—396 (1945).

— Der Bau der Sehnen während verschiedener Altersperioden und unter wechselnden funktionellen Bedingungen. I. Acta anat. (Basel) **6**, 113—140 (1948a).

— The structure of tendons at various ages and under different functional conditions. II. An electron microscopic investigation from white rats. Acta anat. (Basel) **6**, 193—225 (1948b).

— Functionally induced changes in articular cartilage. In: Biomechanical studies of the musculo-skeletal system, p. 3—19 [F. G. EVANS (ed.)]. Springfield (Ill.): Ch. C. Thomas 1961.

KARNER, H. E.: Electron microscope study of developing chick embryo aorta. J. Ultrastruct. Res. **4**, 420—454 (1960).

KRATKY, O., M. LAUER, M. RATZENHOFER, and A. SEKORA: Dependence on age of the x-ray diagram of human tendon collagen. In: Collagen, p. 227—231 [N. RAMANATHAN (ed.)]. New York: Interscience Publ. 1962.

KUHNKE, E.: The fine structure of collagen fibrils as the basis for functioning of tendon tissue. In: Collagen, p. 479—490 [N. RAMANATHAN (ed.)]. New York: Interscience Publ. 1962.

LANG, J.: Über das Verschiebegewebe der Achillessehne. Anat. Anz. **108**, 225—237 (1960).

LINDSAY, W. K., and H. G. THOMSON: Digital flexor tendons: An experimental study. Brit. J. plast. Surg. **12**, 289—316 (1960).

— E. R. TUSTANOFF, and D. C. BIRDSELL: Uptake of tritiated proline in regenerating tendons. Surg. Forum **15**, 459—461 (1964).

LUNDGREN, H. P.: Synthetic fibers made from proteins. Advanc. Protein Chem. **5**, 305—351 (1949).

MACMASTER, P. E.: Tendon and muscle ruptures. J. Bone Jt Surg. **15**, 705—722 (1933).

MASON, M. L., and C. G. SHEARON: The process of tendon repair. An experimental study of tendon structure and tendon graft. Arch. Surg. **25**, 615—692 (1932).

MORGAN, F. R.: The mechanical properties of collagen fibres: stress-strain curves. J. Soc. Leath. Trades Chem. **44**, 170—182 (1960).

MULLER, T.: The effect of estrogen on the loose connective tissue of the albino rat. Anat. Rec. **111**, 355—375 (1951).

NEUBERGER, A., J. C. PERRONE, and H. G. B. SLACK: Relative metabolic inertia of tendon collagen in the rat. Biochem. J. **49**, 199—204 (1951).

—, and H. G. B. SLACK: The metabolism of collagen from liver, bone, skin and tendon in the normal rat. Biochem. J. **53**, 47—52 (1953).

Nisbet, N. W.: Anatomy of the calcaneal tendon of the rabbit. J. Bone Jt Surg. B **42**, 360—366 (1960).

Pahlke, G.: Elektronenmikroskopische Untersuchungen an der Interzellularsubstanz des menschlichen Sehnengewebes. Z. Zellforsch. **39**, 412—430 (1954).

Partington, F. R., and G. C. Wood: The role of non-collagen components in the mechanical behaviour of tendon fibres. Biochim. biophys. Acta (Amst.) **69**, 485—495 (1963).

Peacock, E. E.: Comparison of collagenous tissue surrounding normal and immobilized joints. Surg. Forum **14**, 440—441 (1963).

Pedersen, H. E., F. G. Evans, and H. R. Lissner: Deformation studies of the femur under various loadings and orientations. Anat. Rec. **103**, 159—185 (1949).

Ramachandran, G. N.: Structure of collagen. Nature (Lond.) **177**, 710—711 (1956).

— V. Sasisekharan, and Y. T. Thathachari: Structure of collagen at the molecular level. In: Collagen, p. 81—115 [N. Ramanathan (ed.)]. New York: Interscience Publ. 1962.

Reutervall, O. P. P: son: Über die Elastizität der Gefäßwände und die Methoden ihrer näheren Prüfung. Acta med. scand., Suppl. **2** (1921).

Rich, A., and F. H. C. Crick: The structure of collagen. Nature (Lond.) **176**, 915—916 (1955).

Rigby, B. J.: Effect of cyclic extension on the physical properties of tendon collagen and its possible relation to biological aging of collagen. Nature (Lond.) **202**, 1072—1074 (1964).

— N. Hirai, J. D. Spikes, and H. Eyring: The mechanical properties of rat tail tendon. J. gen. Physiol. **43**, 265—283 (1959).

Robertson, W. van B.: Metabolism of collagen in mammalian tissues. In: Connective tissue — intercellular macromolecules, p. 93—106. Boston: Little, Brown & Co. 1964.

Schmitt, F. O., C. E. Hall, and M. A. Jakus: Electron microscope investigations of the structure of collagen. J. cell. comp. Physiol. **20**, 11—33 (1942).

Schneewind, J. H., F. K. Kline, and C. W. Monsour: The role of paratenon in healing of experimental tendon transplants. J. occup. Med. **6**, 429—436 (1964).

Schwarzacher, H. G.: Untersuchungen über die Skelettmuskelsehnenverbindung. Acta anat. (Basel) **40**, 59—86 (1960).

Sinex, F. M.: Cross-linkage and aging. Advanc. Geront. Res. **1**, 165—180 (1964).

Smith, J. W.: The elastic properties of the anterior cruciate ligament of the rabbit. J. Anat. (Lond.) **88**, 369—380 (1954).

Stanisavljevic, S., and R. Pool, Jr.: The paratendinous apparatus of the digits. J. Bone Jt Surg. B **44**, 910—912 (1962).

Stegemann, H.: Mikrobestimmung von Hydroxyprolin mit Chloramin-T und p-Dimethylaminobenzaldehyd. Hoppe-Seylers Z. physiol. Chem. **311**, 41—45 (1958).

Storey, E.: Relaxation in the pubic symphysis of the mouse during pregnancy and after relaxin administration, with special reference to the behaviour of collagen. J. Path. Bact. **74**, 147—162 (1957).

Stucke, K.: Über das elastische Verhalten der Achillessehne im Belastungsversuch. Langenbecks Arch. klin. Chir. **265**, 579—599 (1950).

— Sehnenbelastung und -ruptur im Tierversuch. Chirurg **22**, 16—18 (1951).

Takashima, K., and N. Tamiya: Studies on collagen biosynthesis with C^{14} as a tracer. Collagen Curr. **5**, 36 (1964).

Thompson, R. C., and J. E. Ballou: Studies of metabolic turnover with tritium as a tracer. V. The predominantly non-dynamic state of body constituents in the rat. J. biol. Chem. **223**, 795—809 (1956).

Tsurufuji, S., and Y. Ogata: Mechanism of the formation of insoluble collagen. Collagen Curr. **5**, 36 (1964).

Verzár, F.: The aging of collagen fibres. Experientia (Basel), Suppl. **4**, 35—41 (1956).

— (1957): Cit. by Hall (1965), p. 392.

— The aging of collagen. Sci. Amer. **208** (4), 104—114 (1963).

Viidik, A.: Biomekaniska studier av rörelseapparatens komponenter. Nord. Med. **72**, 1455 (1964).

VIIDIK, A. and T. LEWIN: Changes in tensile strength characteristics and histology of rabbit ligaments induced by different modes of postmortal storage. Acta orthop. scand. **37**, 141—155 (1966).

— L. SANDQVIST, and M. MÄGI: Influence of postmortal storage on tensile strength characteristics and histology of rabbit ligaments. Acta orthop. scand., Suppl. **79** (1965).

VIS, J. H.: Histological investigations into the attachments of tendons and ligaments to the mammalian skeleton. Proc. kon. ned. Akad. Wet. **60**, 147—157 (1957).

VOGEL, A.: Fine structural characteristics of collagen fibers. Path. et Microbiol. (Basel) **27**, 436—446 (1964).

WALKER, L. B., Jr., E. H. HARRIS, and J. V. BENEDICT: Stress-strain relationship in human cadaveric plantaris tendon: a preliminary study. Med. Electron. Biol. Eng. **2**, 31—38 (1964).

WALLS, E. W.: The microanatomy of muscle. In: Structure and function of muscle. I. Structure, p. 21—61 [G. H. BOURNE (ed.)]. New York: Academic Press 1960.

WERTHEIM, M. G.: Mémoire sur l'élasticité et la cohésion des principaux tissus du corps humain. Ann. Chim. Phys. **21**, 385—414 (1847).

WRIGHT, D. G., and D. C. RENNELS: A study of the elastic properties of plantar fascia. J. Bone Jt Surg. A **46**, 482—492 (1964).

ZARZYCKI, J.: Origin of collagenous fibers in the mammary gland by electron microscope investigations. Folia morph. (Warszawa) **15**, 219—226 (1964).

ZELANDER, T.: Ultrastructure of articular cartilage. Z. Zellforsch. **49**, 720—738 (1959).

Dynamic Considerations in Load Bearing Bones with Special Reference to Osteosynthesis and Articular Cartilage

J. M. Zarek

Introduction

Major osteosynthetic procedures have been practised for a sufficiently long time to allow us to look critically at the results. The analysis at present is usually limited to the examination of the characteristics and the behaviour of the living bone under normal and pathological conditions. This lately has been extended to the analysis of the implant from the biomechanical and metallurgical point of view. In the course of the last twenty years the compatibility of various plastic and metallic materials has been investigated although a criterion for compatibility has yet to be determined. When dealing with load bearing bones this usually has been supplemented by a static stress analysis with assumed values of loads due to gravitational and muscular forces. However, despite all this, clinical experience seems to indicate that success with massive osteosynthetic implants is only moderate, and very often the failures are attributed to mechanical problems. The rather disappointing results with the use of plastics in the Judet type of hip joint reconstruction brought into the picture the consideration of the unexpectedly large forces which that joint transmits. The failures here were of two kinds — either the prosthesis failed mechanically or if the strength of the implant proved satisfactory in many instances a large degree of atrophy of bone led towards a very unsatisfactory situation because of pain and lack of stability.

To prevent future mechanical failures plastic materials in load bearing bones were replaced by metals of a great variety including various types of steels, tantalum, zirconium, titanium and cobalt-chrome-molybdenum alloys. However, the bitter experience of mechanical failures has not ended here and most of the investigators have turned for a solution to simple stress analysis which again brought to life the problem of functional significance of formation of bone and the "trajectorial theory of bone form" in relation to the orientation of trabeculae.

Review of Theories of Formation and Strength of Bone

The formation of bone under normal conditions has been the subject of past investigations, in terms of the function it performs, with the purpose of determining whether stress has any effect on the natural growth of bone or not. In 1867 von Meyer demonstrated sections of the upper part of the femur, and discussed the significance of the trabecular arrangement in it. Culmann, an engineer who was present at the meeting, became interested in the analysis of this inner structure and suggested that the trabeculae, as seen in the frontal plane, were arranged along the lines of the principal stresses produced by a force acting on the head of the femur. He produced a diagram of a Fairbairn crane in which he had computed the lines of principal stresses or the so-called stress trajectories[1]. He based his calculations on the assumption that his crane was made of homogeneous, isotropic solid material resembling the femur in shape and loading. His analysis became the "trajectorial theory" of bone formation, and gave rise to the investigation of the functional significance of the orientation of the trabeculae. However, as the trajectories apply to both compressive and tensile stresses investigators such as von Meyer (1867), Roux (1883), and Wolff (1870) held the view that tensile stresses are responsible for the growth of bone, while Jansen (1920) and Carey (1929) believed that the presence of compressive stresses stimulates the formation of bone. Triepel (1922), however, completely rejected the validity of the trajectorial orthogonality of the trabeculae in the bones. Koch's (1917) analysis of the functional significance of bone form supports the view that the trabeculae arrange themselves along the lines of principal stresses for the neck and the head of the femur. He produced calculated results of principal stresses for the neck and the head of the femur when the head carries a concentrated load of 100 pounds. But an examination of his results suggests that they are not in agreement with either the theory of stress analysis or the assumptions he had made for his analysis.

Modern surgical procedures in the reconstruction of the upper part of the femur have stimulated a new interest in the functional significance of bone form and this has been receiving a great deal of attention both from orthopaedic surgeons and engineers. The subject was discussed by specialists in both fields at a Symposium on Biomechanics sponsored by the Institution of Mechanical Engineers in London (1959), and in orthopaedic scientific journals by Haboush (1953), Tobin (1955), Smyth et al. (1964), Bingold (1959), and Garden (1964). Large numbers of these eminent surgeons, observing that without force there is no architecture, use this as a working hypothesis to justify the acceptance of the belief that the macroscopic behaviour of load carrying bone conforms with the principle of "Wolff's law."

It is unfortunate that Culmann's suggestion of the trajectorial theory has received so much attention as it can be criticised on the following points. His assumption that bone is homogeneous is not valid and his analysis is based on two-dimensional stresses, although three-dimensional analysis is necessary. Further his analysis refers to static conditions of loading in the frontal plane while the actual force on the head of the femur is mostly of a dynamic character with its directions of action oscillating appreciably, so that the maximum value of the force does not necessarily

---

[1] It can be shown that in a most general state of stress at a point in an elastic material there will always be three mutually perpendicular planes on which the shearing stresses vanish and the direct stresses have stationary (maximum or minimum) values. These direct stresses are called the principal stresses and their action may be tensile or compressive. In a two-dimensional or a plane system of stress there will be two mutually perpendicular planes on which the shearing stresses will vanish and the direct stresses will have maximum and minimum values. The directions of these direct or principal stresses will have maximum and minimum values. The directions of these direct or principal stresses which are at a point at right angles to each other are called the principal directions. It is possible to construct two systems of orthogonal curves the tangents of which coincide with the directions of the principal stresses at each point. These curves are called "stress trajectories."

coincide with the frontal plane. The frontal cross-sectional plane, used by Culmann for his analysis, need not be one of the principal planes and the directions of the two principal stresses need not be parallel to the frontal plane. In addition, the intensity of the stress along each trajectory is greatest where it is parallel to the longitudinal axis of the neck of the femur and diminishes to zero at the point where the trajectory cuts the outer fibre of the bone at right angles. In view of this, it is difficult to accept the opinions of those who believe in the mechanical adaptation of bone, that is, that the trabeculae lie along the paths of maximum stress within the bone and thus transmit a maximum load with a minimum of material. Other explanations must be sought.

Dynamic Loading of Bones

It is true to say that the "static" stress analysis of bone has helped to clarify the picture to a certain extent only as the results depend largely on the assumptions made for the loading or the nature of the forces acting on the bone. The forces acting, for instance, on the femur consist of:

1) gravitational forces,
2) acceleration and deceleration forces,
3) forces due to the action of muscles.

All these forces frequently act in a dynamic manner and very often create conditions similar to those experienced under shock-loading, where, from simple considerations of work or energy, it can be shown that the stress reached under conditions of impact may be twice the static stress produced by the same load if applied in a static manner. Present-day literature dealing with the problem of fractures takes into consideration qualitatively the existence of dynamic conditions in load bearing bones. However the final analysis is always reduced to static conditions and conclusions with regard to failure are drawn in terms of the maximum tensile stress developed in the bone or the prosthesis.

It is reasonable to consider that in a simple tensile system of loading, the elastic limit is directly linked with a certain value of tensile stress; but it must be remembered that during the action of the tensile stress such quantities as shear stress and strain energy also reach a definite value and therefore any one of these may be a deciding factor in the mechanical failure. In a most general complex system of stress, the maximum shear stress and the elastic energy in the materials can be evaluated in terms of the known stress components and the elastic constants of the material. The difficulty lies in the choice of which quantity causes the material to pass beyond its elastic limit or, in other words, is the criterion of failure. In engineering practice having decided on a quantity as a criterion of failure, the actual value of that quantity which corresponds to the beginning of failure is usually taken to be the value it reaches in the simple tension case at the elastic limit. In the case of brittle materials usually the maximum principal stress criterion is used while for ductile materials the maximum shear stress, strain energy, especially shear strain energy, criteria give satisfactory results[1].

---

[1] For more detailed description of the various theories of failure the reader is referred to engineering textbooks on stress analysis.

However, bone as an engineering material is not homogeneous and shows a marked degree of anisotropy which indicates that the elastic properties of bone are not the same in all directions. Looking, for instance, at the femur as a complete engineering structural element, from the experiments on fracturing femurs under various load conditions in the laboratory and the analysis of these fractures as they normally occur in life, it appears that the structure of the neck of the femur and the complete shape of the femur are so arranged as to act as a "shock-absorber" and be capable of absorbing a great amount of elastic energy under dynamic load conditions. Therefore, such characteristics as load-deflection curves must be considered of great significance in the analysis of the load bearing capacity. In past studies of prostheses the energy aspect of the problem has been badly overlooked. LISSNER and EVANS (1956) and EVANS et al. (1958), in the course of their extensive biomechanical studies of the various parts of the human skeletal system, expressed the view that any biomechanical considerations of the skeleton would be incomplete unless they included investigations of the energy to produce failure of the cortical bone. Their study of the fractures of the skull and the pelvis under dynamic conditions is worthy of noting as it demonstrates that these bones due to their shape represent excellent energy absorbing structures. The case of the femur, the skull, and the pelvis are the more obvious examples illustrating the idea that the skeletal system is designed as an energy absorbing unit and further progress towards successful osteosynthesis will largely depend on our appreciation of the importance of this point.

ZAREK and EDWARDS (1964) extended the energy absorbing idea to the functional behaviour of articular cartilage in load bearing joints. A number of authors have suggested that articular cartilage probably acts as a shock-absorber; however, little evidence has been provided to substantiate this view and no one has given a detailed explanation of how cartilage functions when it is dynamically loaded. This is probably because a satisfactory understanding of even the most fundamental aspects of cartilage mechanics has been achieved only in the last few years and the more complex aspects of the problem have yet to be solved.

Mechanical Structure and Behaviour of Cartilage

The structure of articular cartilage is far from resembling engineering materials such as metals to which the application of standard theories of stress analysis gives very satisfactory results. The application of the engineering stress analysis is, therefore, somehow limited and it is imperative that these limitations should be understood.

Articular cartilage appears in joints as a thin, elastic covering over the osteoarticular surfaces and so prevents the opposed bony surfaces from making a direct contact. For the purpose of mechanical analysis, uncalcified articular cartilage may be taken to consist of cells distributed at random throughout a greater quantity of matrix of collagen fibrils surrounded by a formless ground substance. The latter is attached to the fibrils as a fine granular coating and together they form a three-dimensional network of interlacing fibres with a continuous system of pathways or pores between the network. These pores are very small but sufficiently well established to allow permeation of liquid through the cartilaginous matrix at a very low rate. The synovial fluid in which the cartilage is normally bathed is a highly viscous fluid with non-Newtonian properties. When a joint is functioning, the articular cartilages are alternatively compressed and released. During the release phase, when the compressive load is decreasing, the cartilages expand and assimilate liquid from the free synovial fluid in the joint cavity, and from the underlying marrow cavities.

The presence of this pore liquid in the cartilage has a most pronounced effect on the way in which it supports load. If cartilage were devoid of liquid, any load applied to it would almost instantaneously induce elastic stresses in the cartilaginous matrix and cells. However, when cartilage is saturated with liquid, the application of an external load causes excess fluid pressures to be developed in the pores. These fluid pressures carry part of the load, leaving only the remaining part to be supported by compressive stresses in the solid material. The pressure of the liquid in the voids at the free articular surface must always equal the environmental pressure in the joint cavity, which will be less than the excess fluid pressures in the deeper regions of the cartilage. Thus the pore fluid will flow from the deeper regions of high pressure towards the superficial regions of low pressure, gradually causing fluid to be squeezed from the cartilage.

In applied sciences, a process involving a decrease in the liquid content of a porous material, when an external load is applied to it is called "consolidation." It is most marked in materials possessing high compressibility and high resistance to the flow of liquid through its interstitial spaces. Articular cartilage displays both these properties. The basic principle of the mechanism of consolidation in articular cartilage may be explained by reference to the classical analogy consisting of a cylindrical container fitted with a number of pistons which are separated by springs. The pistons are drilled with small holes through which the liquid filling the container can flow against a high resistance. Detailed explanations of the analogy are given in a previous paper (Zarek and Edwards, 1964) to which the reader is referred. Those who are not familiar with the engineering literature and who wish to inquire into the relationship between stresses and hydrostatic pressure are referred to the theory of consolidation for porous materials saturated with liquid, published by Skempton (1960) who gives the formula

$$\bar{p}=p_1-cu$$
$$\bar{p}=p_2-cu$$
$$\bar{p}=p_3-cu$$

in which $\bar{p}_1$, $\bar{p}_2$, and $\bar{p}_3$ are the stresses developed in the matrix and solid cellular constituents which in the principal directions of the applied stresses; p_1, p_2, and p_3 are the applied principal stresses; u is the hydrostatic pressure in the pore liquid; and c is a coefficient depending on the consolidation properties of the material.

In the process of consolidation, loss of liquid from the cartilage must permit the fibrous matrix to deform and carry increasing proportions of the applied load. Therefore, the magnitudes of the liquid pressures and the matrix stresses developed at a particular instant of time will depend upon the degree of consolidation attained.

Under dynamic loading conditions the stresses developed for a certain load value are much higher than in a case of static loading. While the mechanics of a cartilage subjected to a static load may be explained in terms of static equilibrium of forces, to appreciate fully the effect of dynamic loading on the action of articular cartilage it is necessary to take energy relations into consideration.

ZAREK and EDWARDS (1964) illustrated the essential details of the energy absorbing process by using the previously mentioned piston-spring analogy. Considering that the magnitude of the load applied suddenly is sufficiently large to cause accelerated flow of the pore liquid they gave an account of how articular cartilage minimizes the load it transmits by dissipating energy. When a load is applied dynamically, its motion is retarded by the cartilage until it comes to rest at the static equilibrium position. The resultant external force, which acts on the cartilage at any time during this dynamic process, equals the static weight of the load plus the inertia force due to its retardation. It is balanced and transmitted to the subchondral bone by the resilience of the matrix and the pressures in the pore liquid. Differences in liquid pressure from one part of the cartilage to another give rise to a flow of liquid which dissipates part of the energy supplied by the load. Therefore, the maximum force transmitted to the bone is smaller than it would have been if there were no cartilages in the joint. In this way, articular cartilage minimizes the stresses to which bone is subjected by absorbing energy.

If the energy absorbing capacity of all the articular cartilages in the body were to be greatly reduced, failure of one or more of the skeletal segments would not necessarily result as soon as an action involving large dynamic forces had to be performed. Many of the bones are constructed in a way which allows them to withstand the application of appreciable amounts of energy before failure occurs.

Osteosynthesis of Load Bearing Bones

The Bone

In the reconstruction of bones, it must be remembered that all load bearing bones are subjected to the action of mechanical forces that are usually dynamic in character. The femur, of all the bones of the skeleton, has been studied most extensively, owing to its frequent fractures and to its liability to arthritis at the hip joint. Under the different conditions met, the femur has been nailed, screwed at the joints of fracture or, in the case of pathological changes, parts of it have been removed and replaced by plastic or metallic implants.

The femur as a structural element represents a strut with some initial curvature and as such, when loaded in its longitudinal direction in addition to other effects, is also subjected to bending. Load-deflection characteristics, therefore, are of great importance in assessing the energy absorbing capacity of the femur. Replacement of a large part like the head and the neck by metallic implants usually disturbs the general natural behaviour of the femur. Firstly, it causes a large degree of re-orientation of forces in the bone from their natural anatomical direction. Secondly, it introduces local intensities of stress far in excess of these experienced under normal conditions. Thirdly, the load-deflection relationship because of the different rigidity of the implant undergoes changes with the inevitable introduction of high degree of load localization.

To assess the degree of energy absorption tests were carried out on two similar femurs. One was subjected to dynamic loading in a manner similar to conditions experienced in life with the femur as a whole structure taking the load. The second one (from the same subject) was mounted in the testing machine in such a manner that only the upper part was subjected to the impact as the lower part was rigidly mounted about 2 in. below the greater trochanter.

It was found that the second femur failed when 0.6 of the load which caused the failure of the neck of the femur in the first case was applied. It resulted in the fracture and collapse of the cortical layer of the head — although in both experiments the same acetabulum-like cup lined with a thin layer of rubber was used.

The fixation by means of plates and screws very often produces problems because of bent plates or more often because of broken screws. It is common practice, when using screws in the plates, to fix them in one plane, that is, usually in the plane of the plate radially to the centre of the shaft of the bone. However, the load which the screws are supposed to carry varies not only in magnitude but also in direction. To allow for this directional oscillation the screws should not all be inserted radially but the angle should vary. Also greater stability would be obtained if, instead of inserting the screws in the normal cross-sectional plane of the shaft, oblique planes with directions of the screws towards the fracture which is to be stabilized were used. This arrangement would contribute not only a more stable fixation but would also give a much more favourable field of stress distribution in the screws and the bone.

To continue with the femur — let us consider some of the problems involved in the repair of the subcapital fractures of the femur which, thanks to the active interest of GARDEN (1964) and SMYTH et al. (1964), both orthopaedic surgeons, has stimulated a great deal of further thought. GARDEN (1964) says "Many surgeons are now convinced that the 'unsolved' fracture should be renamed the 'unsolvable' fracture and the defeatist attitude of Sir Astley Cooper still lingers in present-day practice. This is reflected by the increasing tendency to abandon treatment by reduction and fixation, and to replace the femoral head with a prosthesis. This policy, which amounts to a confession of failure, would be fully justified if every subcapital fracture failed to unite. But non-union does not always occur, and many such fractures heal with modern methods of treatment. There must be a scientific explanation for the fact that union occurs in some, but not in all, subcapital fractures, and it would seem more logical to search for this explanation than to accept the widespread belief that there is something unfathomable about these injuries." He finally concludes by quoting WATSON-JONES who says "A perfect result may be expected from a technically perfect operation; an imperfect result is due to imperfect technique." Here, therefore, we may ask what do we understand by perfection of operative technique when on the path to a foolproof method of fixation of the displaced subcapital fracture many questions remain unanswered.

GARDEN (1964), in the course of his analysis, came to the conclusion that a stable reduction in the subcapital fractures of the femur is obtained when the muscular and gravitational forces tending to redisplace the fracture are opposed by equal and opposite counter-forces. His clinical solution to the problem lies, therefore, in the fixation by two crossed screws one of which is inserted through the anterior half of the greater trochanter towards the inferior part of the femoral head while the other is positioned so that it passes through the lateral femoral cortex just above the calcar femorale.

SMYTH et al. (1964), great believers in, and students of, the functional adaptation of bone, described their operative technique with a set of two crossed screws in a position similar to that used by GARDEN (1964) but interconnected with a plate. They explain, by very sound reasoning, the mechanical advantages from such triangular fixation towards achieving a better stability which in turn leads to a more successful union.

GARDEN (1964) and SMYTH et al. (1964) claim great improvement in their clinical results and their analysis is based on sound mechanical principles. SMYTH et al. (1964), great students of the functional architectural structure of the neck of the femur, re-emphasize the statement of EVANS et al. (1951) that the greater trochanter has an excellent energy absorbing capacity because of the spongy cancellous bone underlying its elastic cortex. From the point of view of stress analysis, it must be added that, the greater trochanter, situated in a position where the bending moment in the femur appears to be greatest prevents high stress concentrations. Anyone acquainted with analysis of stress in structural elements — examining the shape and the loading of the femur under normal conditions — will conclude that it has an exceptionally uniform field of stress. The smaller trochanter may be used here as a less obvious example of this, being situated in the position of an engineering fillet to prevent undesirable stress concentration. Hence EVANS et al. (1951), in the course of their detailed experimental studies, demonstrated the large load carrying capacity of the upper part of the femur and failures occurring due to tensile stresses. The stresscoat pattern of cracks which they obtained, however, demonstrated clearly that no high degree of stress concentration occurs and that the field of stress is relatively uniform, i.e., free of discontinuities.

From the above comments, it seems indicative that simple stress analysis, in terms of tensile and compressive stresses, represents only a partial tool towards the solution of problems associated with the load bearing characteristics of the skeletal system. The ultimate magnitude or peak intensity of stress appears to be of lesser importance under everyday dynamic conditions than the sudden changes in the magnitude of the field of stress giving rise to some degree of stress concentration. Perhaps further work, which would relate the behaviour of, say, such elements as the femur to the various engineering theories of failure might contribute towards a more reliable criterion of failure. A more detailed quantitative analysis of the energy absorbing characteristics of the skeletal system will no doubt increase the accuracy of calculations in the course of reconstructive osteosynthesis.

The Implant

The requirements which compatibility and static stress analysis impose on an acceptable prosthesis have been discussed at length in past literature. The various characteristics, however, were stressed in the light of the picture of the problem as seen at that time. The present analysis indicates that search for materials capable of giving load bearing characteristics similar to bone must still continue. The cobalt-chrome-molybdenum alloys commercially available in a cast form are clinically quite acceptable. However, impurities and, in the case of screws, large grain size contribute very often to failure. Recent development of a sintered cobalt-chrome-molybdenum alloy (ZAREK et al., 1946) has improved characteristics, but, its commercial availability still awaits solution.

With regard to design, all prostheses should be free of any stress concentration and should transmit load to the bone as uniformly as possible. In addition, the re-orientation of forces from the natural anatomical direction should be as small as possible and the deflection curves should approach those of the bone.

Fig. 1 shows diagrammatically a massive prosthetic replacement at the upper part of the femur. No screws were used and the method of fixation against bending relied on an intramedullary stem and a cup-like cover over the end of the shaft of the femur. Rotation was prevented by suitably shaping the outside of the cortical

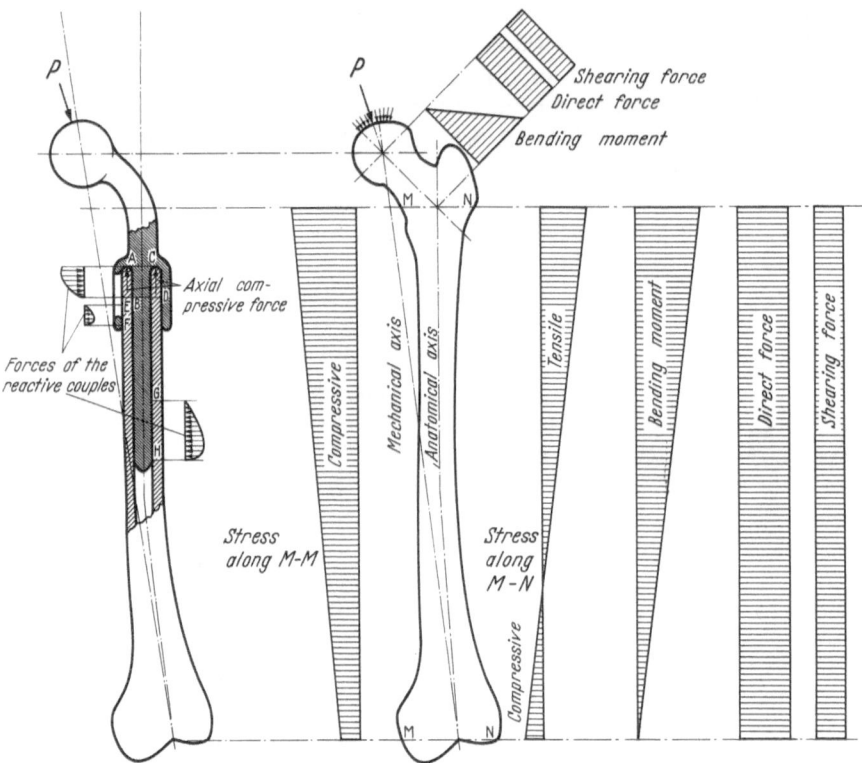

Fig. 1. The right-hand part shows diagrammatically the forces and stresses present in the femur in the frontal plane when it is subjected to a load P acting in that plane. The left-hand diagram illustrates the distribution of forces acting on the shaft of the femur when the upper part of it is replaced by a prosthesis (J. M. Zarek, Modern Trends in Surgical Materials, p. 113. Butterworth & Co. Ltd. 1958. Reproduced by permission of the publisher)

end of the bone and the cup. It may be noted here that during walking the largest part of the load is taken by the axial compressive force acting on the ring of the cortical bone at A and C. The bending moment produces certain reactive couples, which, however, are conveniently distributed so as not to give rise to points of high stress concentration on the bone and cause absorption, because they act in directions which do not coincide with the anatomical action line of forces in the bone. The intramedullary nail was made as long as possible because the longer the nail the smaller the reactive forces of the couples. To avoid any stress concentration effects the changes in the cross section of the prosthesis were made gradually. The cup had three large slots which appear to be essential in situations of this nature to prevent stagnation conditions for body fluids at the end of the shaft of the bone.

Fig. 2 shows that a few months after the operation new cortical bone was forming very readily along the medial part of the cup which coincides with the region in which a compressive force acts on the bone. The described prosthesis was used in a patient in 1954 and still gives very satisfactory service.

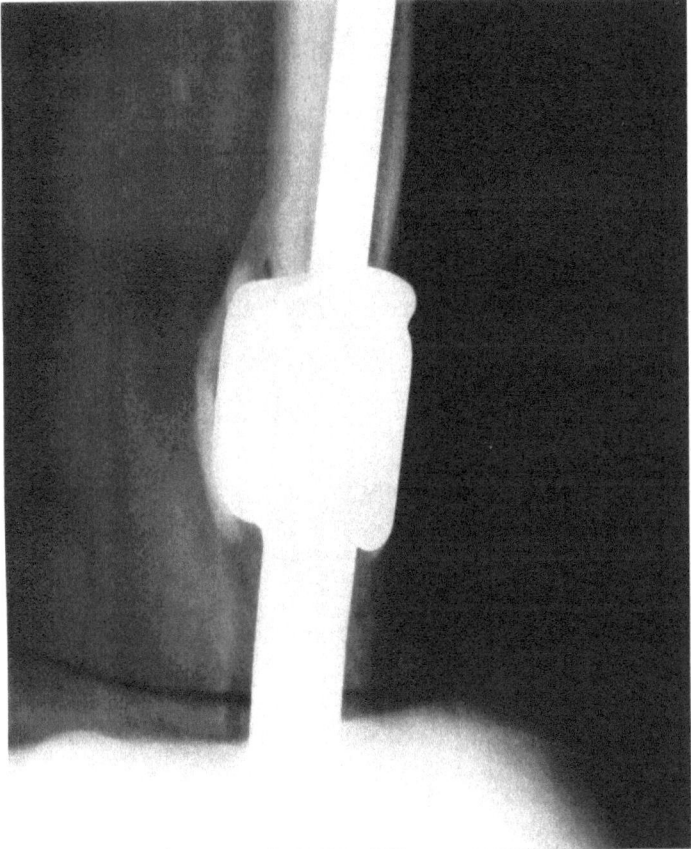

Fig. 2. Replacement of the upper part of a femur showing good bone formation over the prosthesis a few months after insertion (by courtesy of A. C. Bingold, F.R.C.S.)

Fig. 3 illustrates a close-up view of the "eye" in the upper part of the stem of a Moore's prosthesis which shows that new bone has bridged over two sides of the cortical bone through the "eye" thus contributing to a very stable "assembly."

Illustrations 2 and 3 indicate that to promote growth of bone at the junction with an implant stagnation conditions for body fluids must be avoided.

Finally — one comment about the orthopaedic screw. A semi-circular groove. instead of the industrial V-notch, would reduce appreciably the number of failures,

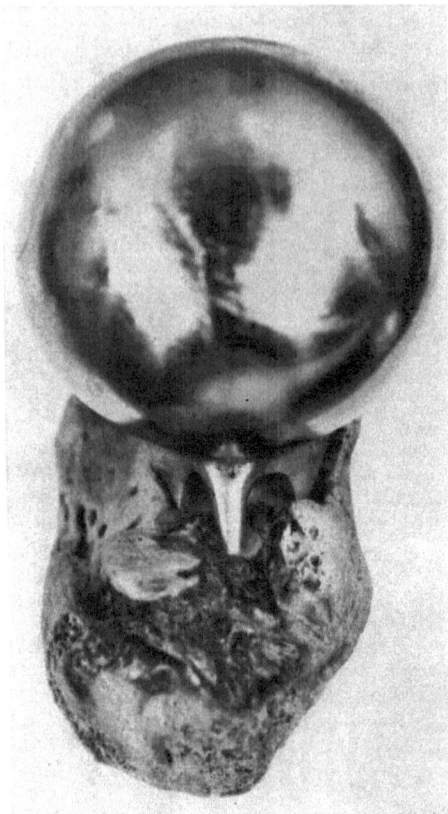

Fig. 3. Top view of a Moore's prosthesis showing bone formation through the "eye" of the stem

Summary

1. In view of the renewed interest in the functional adaptation of bone the various theories are briefly reviewed with comments.

2. The static stress analysis appears to be inadequate in the load bearing osteosynthesis.

3. Suggestion is made that an investigation into the engineering theories of elastic failure, as applied to bone, may be helpful towards a better understanding of requirements for successful osteosynthesis.

4. Dynamic loading conditions of bone suggest that any biomechanical considerations of the skeletal system would be incomplete unless they include investigations into the level of energy required to produce failure of the cortical bone.

5. The idea that the skeletal system represents a very efficient shock absorbing system is discussed and examples are given.

6. It is suggested that the articular cartilage represents a very efficient shock absorbing element within the skeletal system. This is followed by relatively brief descriptions of the mechanical structure and behaviour of cartilage. The engineering theory of consolidation is used to explain the mechanism of how cartilage supports load and absorbs energy.

7. Because of the importance of energy some of the problems of osteosynthesis in load bearing bones are analysed. Firstly, the behaviour of bone under dynamic conditions is discussed and secondly, the requirements which an implant should satisfy to give a greater degree of clinical success. Two examples illustrate some of the points discussed.

Acknowledgement

Thanks are due Mr. J. EDWARDS whose assistance in this work is much appreciated.

References

BINGOLD, A. C.: Experimental work on femoral neck fractures. Proc. roy. Soc. Med. **52**, 906—910 (1959).

CAREY, E. J.: Studies in the dynamics of histogenesis. Radiology **13**, 127—168 (1929).

EVANS, F. G., H. R. LISSNER, and M. LEBOW: The relation of energy, velocity, and acceleration to skull deformation and fracture. Surg. Gynec. Obstet. **107**, 593—601 (1958).

— H. E. PEDERSEN, and H. R. LISSNER: The role of tensile stress in the mechanism of femoral fractures. J. Bone Jt Surg. A **33** (2), 485—501 (1951).

GARDEN, R. S.: Stability and union in subcapital fractures of the femur. J. Bone Jt Surg. B **46**, 630—647 (1964).

HABOUSH, E. J.: A new operation of arthroplasty of the hip based on biomechanics, photoelasticity, fast-setting dental acrylic and other considerations. Bull. Hosp. Jt Dis. (N.Y.) **14**, 242—277 (1953).

JANSEN, M.: On bone formation, its relation to tension and pressure, p. 114. Manchester: University Press; London: Longmans, Green & Co. 1920.

KOCH, J. C.: The laws of bone architecture. Amer. J. Anat. **21**, 177—298 (1917).

LISSNER, H. R., and F. G. EVANS: Engineering aspects of fractures. Clin. Orthop. **8**, 310—322 (1956).

MEYER, G. H.: Die Architectur der Spongiosa. Arch. Anat. Physiol. wiss. Med. **34**, 615—628 (1867).

ROUX, W.: Beiträge zur Morphologie der functionellen Anpassung. 2. Ueber die Selbstregulation der morphologischen Länge der Skeletmuskeln. 70 pp., 8⁰ Jena: Gustav Fischer 1883 from Jena. Z. Med. Naturw., N.F. **16** (1882/83).

SKEMPTON, A. W.: Pore pressure and suction in soils. The Internat. Soc. of Soil Mechanics. Conference organised by the British National Society of Soil Mechanics and Foundation Engineering in March 1960. London: Butterworths & Co. 1961.

SMYTH, E. H. J., J. S. ELLIS, M. C. MANIFOLD, and P. R. DEWEY: Triangle pinning for fracture of the femoral neck. J. Bone Jt Surg. B **46**, 664—673 (1964).

Symposium on Biomechanics. Institution of Mechanical Engineers, London 1959.

TOBIN, W. J.: The internal architecture of the femur and its clinical significance: the upper end. J. Bone Jt Surg. A **37**, 57—72, 77, 88 (1955).

TRIEPEL: Die Architektur der Knochenspongiosa in neuer Auffassung. Z. menschl. Vererb.-u. Konstit.-Lehre **8**, 269 (1928).

WOLFF, J.: Ueber die innere Architectur der Knochen und ihre Bedeutung für die Frage vom Knochenwachstumh. Virchows Arch. path. Anat. **50**, 389—453 (1870).

ZAREK, J. M., and J. EDWARDS: Dynamic considerations of the human skeletal system. Symposium on Biomechanics and related Bio-engineering topics. The University of Strathclyde, Glasgow. Oxford, Pergamon Press Ltd. 1964.

— A. S. SMITH, and A. E. F. WILKINSON: Sintering cobalt-chrome-molybdenum alloys for osteosynthesis. Nature (Lond.) **203**, 900 (4947) (1964).

Intravital Measurements of Forces Acting on the Hip-Joint

N. Rydell

The design of the proximal femur, i.e., the neck and the head, and its internal architecture is considered to be related to mechanical stress and strain. The importance and frequency of clinical conditions of this region, in addition to the immediate suggestion of a load bearing construction offered by its anatomy, are the probable reasons for the great attention given to analysis of the forces in the hip-joint and to the determination of the force acting on the femoral head.

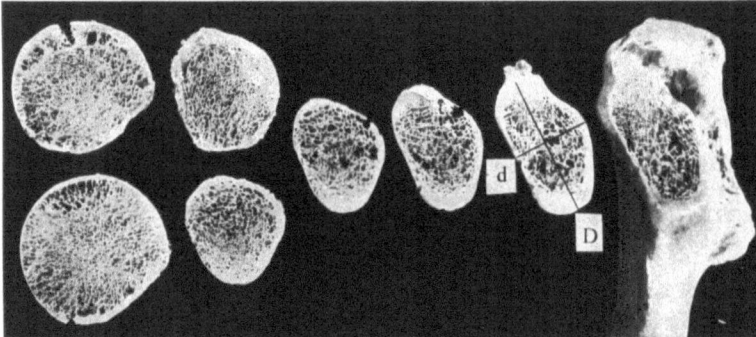

Fig. 1. Cross sections through the femoral neck. In the lateral cross section the major axis D and the minor d are marked. Note their direction in relation to the shaft. (Slightly modified from BACKMAN, 1957)

The hip-joint is a ball and socket joint. The femoral head is not a perfect sphere but is slightly compressed in an approximately ventro-dorsal direction. The difference between the two principal axes is, however, very small, the ratio between them being 1:02.

The femoral neck extends from the head in a distal-lateral direction. It is 30—40 mm long and, like the head, compressed in an approximate ventro-dorsal direction. The shape of the neck changes in its course. At its junction with the head a cross section through the neck is, like the head, almost spherical (Fig. 1). The elliptical form of the neck increases in distal-lateral direction. In the middle of the neck the ratio between the two principal axes is 1:18 and at the junction between the neck and the femoral shaft the ratio is 1:65 (BACKMAN, 1957).

The major axis of the elliptical cross section is not in the same direction as the long axis of the femoral shaft but forms an angle opening dorsally with relation to the shaft. The cortical shell of the neck is thin as paper close to the head but distally it gradually increases in thickness. The increase is small in the upper part of the neck but the thickness increases considerably in the inferior part and attains its maximum where the major axis intersects the cortical bone. In a cross section of this area the cortical bone is, therefore, thickest in the inferior part in a plane through the major axis (Fig. 2). It is interesting to note that a newborn baby has a circular femoral neck and that the elliptical form develops during growth.

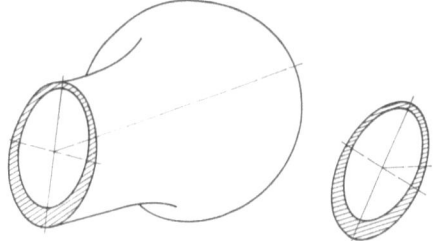

In order to understand relations between the design of the upper femur and forces acting on the bone the angles between the neck and the femoral shaft must be known. This presupposes that the axes and the planes which are used for determinations of the angles are defined. Fig. 3 shows the axes and the planes according to BILLING (1954) and BACKMAN (1957). Definitions of the angles used in this study are given below.

Fig. 2. A cross section through the femoral neck showing thickening of the cortical bone in the inferior part of the neck in the plane of the major axis. (Modified from BACKMAN, 1957)

The angle the neck forms with the femoral shaft, the *cervico-diaphysial angle*, is defined as the angle between the cervical axis, *OHC*, and the ideal shaft axis, *AOK* (Fig. 3). The magnitude of the angle is, on the average, 126⁰.

The cervical axis, *OH*, usually deviates forward to the frontal plane of the femur, *ABK*, and forms with this plane an angle *v*. A plane through the ideal axis and the cervical axis *AOKC* is called the *antetorsion* or the *anteversion plane*. The magnitude of antetorsion is given by the angle between the frontal plane of the femur *AOKB* and the antetorsion plane *AOKC*, the angle *t*. This angle is, on the average, 14⁰ but the variation is great (BACKMAN, 1957).

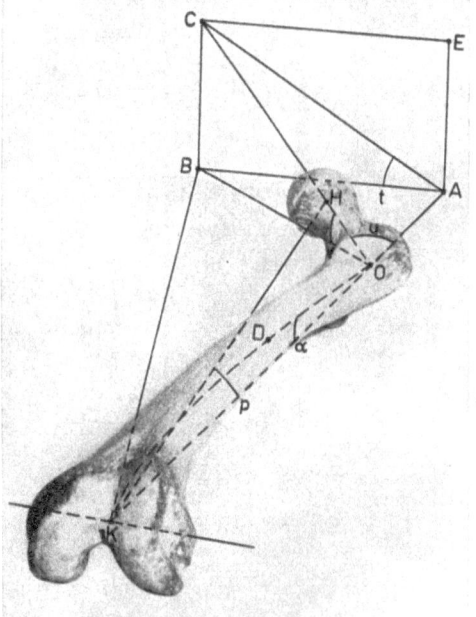

Fig. 3. The planes and axes of the femur used in this study. (From BACKMAN, 1957)

If a plane is laid through the head and the neck of the femur, coinciding with the cervical axis and the major axis of the section area (Fig. 4), it will run through the middle of the fovea and bisecting divide the head and the neck into two symmetrical halves. This plane, called the *principal plane*, coincides with the major axis of the section area and must form an angle opening dorsally with the cervico-diaphysial plane. This angle is, on an average, 25⁰.

The variations of the cervico-diaphysial angle, the antetorsion angle, and the principal plane under normal and pathological conditions have been the subject of different mechanical interpretations. Some think that changes can take place in the design of bone as a result of adaptation to resist existing mechanical conditions, while others claim that changes occur from yielding to mechanical stresses. Genetic factors also play a part.

Studies of the literature give a puzzling impression. Most authors apparently do not distinguish between functional adaptation and yielding. For example, the fact that the cervico-diaphysial angle is 150⁰ at birth and decreases to 126⁰ during adolescence is supposed to be the result of a less vertical load. Paralysis of the

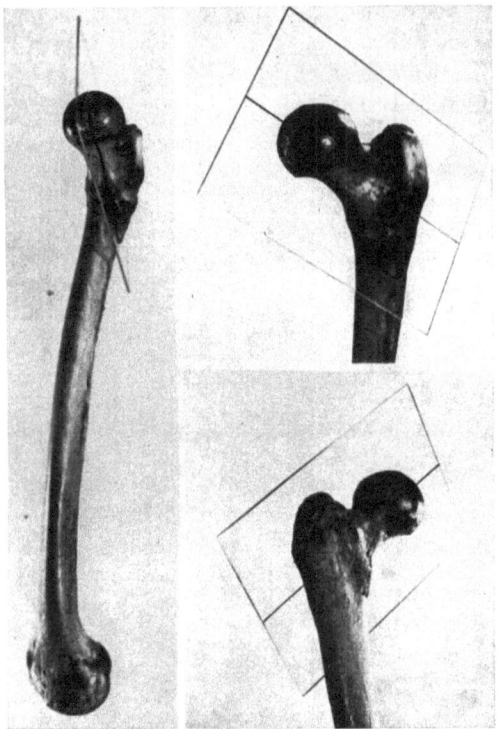

Fig. 4. Principal plane of the femoral head and neck. (From BACKMAN, 1957)

abductor muscles in children results in a coxa valga because of vertical loading on the hip-joint. These changes may be the result of adaptation to mechanical conditions. The antetorsion or anteversion of the neck starts during uterine life and reaches its greatest magnitude at birth after which it decreases to its final value. As the deviation is probably because of torsion of the upper femur, the word *ante-torsion* is preferable. The change of the antetorsion angle is usually regarded as a yield to mechanical stresses.

The elliptical shape of the cross-sectional area of the femoral neck gives a high resistance against vertical forces. As the elliptical form is increased distally the neck is a body of almost equal strength in all directions, for the load applied, which means that stresses are equal throughout its length. The fact that the major axis of the ellipse forms an angle opening dorsally with the femoral shaft has long been known. BACKMAN (1957) concluded that the inclination of the principal plane means that the force acting on the hip-joint must, under the most important conditions, attack the femoral head from its ventral side disputing the direction of the force suggested by PAUWELS (1935). The fact that the cortical bone is thickest on the inferior side of the neck in the principal plane supports BACKMAN's assumption.

The internal structure of the proximal femur has also been the subject of mechanical analysis. The spongiosa forms a trabecular network in which three main systems

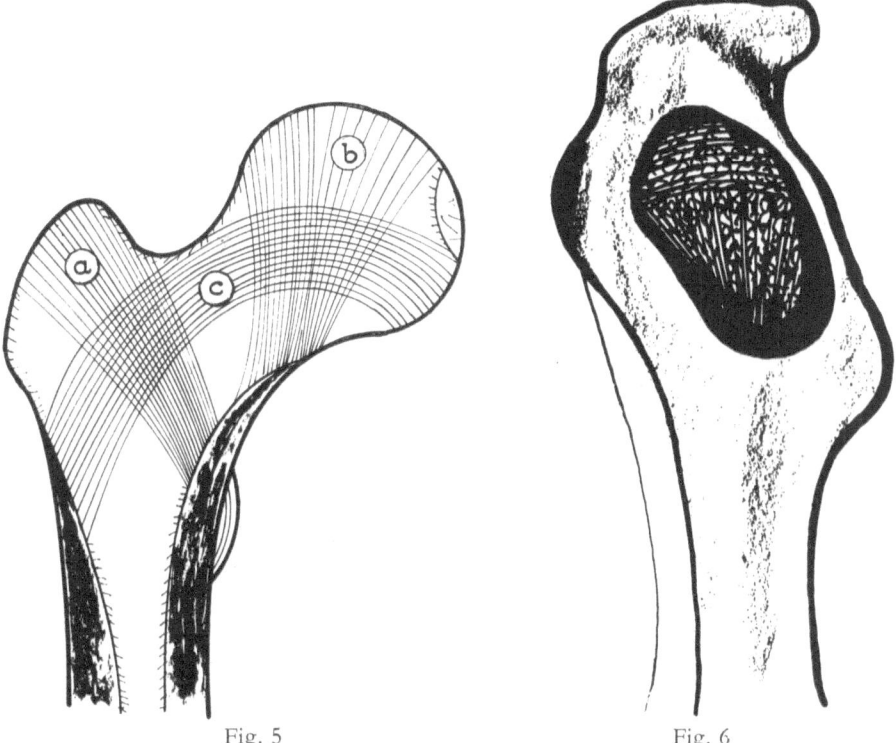

Fig. 5   Fig. 6

Fig. 5a—c. The trabecular systems of the upper part of the femur: a, the lateral system, b, the medial system, c, the arcuate system

Fig. 6. A cross section through the femoral neck near its junction with the shaft showing the T-shaped trabecular network in this region

can be identified: the medial, the lateral, and the arcuate (Fig. 5). BOURGERY (1832) described the internal architecture of the femoral head and neck and claimed that the trabecular system increases the strength of the bone. He felt that the medial trabecular system carries the load acting on the head through the neck to the cortical layer of the femoral shaft. He also described the two triangles formed by the different trabecular systems, the medial one later being named Ward's triangle.

The function of the trabecular systems is still discussed. Usually the medial part is, as BOURGERY (1832) stated, thought to take compressive stress. The function of the arcuate portion is unclear. An interesting thing is, however, that in a cross section through the neck, close to the shaft, the lamellae of the spongiosa are arranged in the form of a T (Fig. 6). This orientation could mean that the spongiosa at this level acts as a T-beam to give highest resistance against vertical loading. The base of the T comes from the lateral trabecular system and the upper part of the T from the arcuate part. Forming a T-beam to resist vertical loading could be one explanation of the function of the two systems.

In 1866 Meyer and Culmann (Meyer, 1867) pointed out that the trabeculae in the femoral head and neck resemble the stress-trajectories in a loaded Fairbairn crane.

In the past photoelastic model tests have been performed in studying stress distribution in the upper femur. There are some disadvantages with these tests because they are only valid for one plane and the exact nature of all physiologic loads is not known.

It is interesting to note that higher primates have a less well developed trabecular system and that the Neanderthal man found at Spy in Belgium lacks Ward's triangle. Animals, e.g., sloths, that spend most of their lives hanging in the trees, lack the trabecular system in the proximal femur (Smyth, 1958).

Although the literature on mechanical interpretations of femoral design and internal structure is puzzling, Wolff's publications (1870—1900) must be mentioned because they led to the formulation of Wolff's law of bone architecture. Wolff's law states, in short, that in bone there is a transformation force that produces a functional bone design. Modifications in the form and function of a bone are followed by changes in its internal architecture and secondary alterations in its external configuration in accordance with mathematical laws. Roux (1893 and 1895) introduced the concept of "functional adaptation." Doubtless, some problems could be clarified if the magnitude and direction of the force acting on the hip-joint during different conditions could be determined.

Many experiments have been made to produce fractures of the femoral neck, often by submitting the head to a vertical force. However, the fracture thus produced is not a clinical type fracture. This phenomenon was explained, on a mechanical and anatomic basis, by Bourgery (1832) who stated that when a vertical load coming from above is transferred by the medial trabecular system to the femoral shaft, a fracture is not likely to occur. If the load comes from beneath or, still worse, from the trochanteric area, the medial trabecular system will not transfer the load and the two triangle-shaped spaces (Fig. 5) will be submitted to stress. As the density of the bone is minimal in these spaces, fractures are likely to occur. If the fracture starts through the inner triangular space, an intracapsular fracture of the neck appears, whereas a fracture through the lateral triangle produces an extracapsular or an isolated trochanteric fracture.

For the design of an osteosynthetic prosthesis and knowledge of how to place it, it is of greatest importance to know the magnitude and direction of the force which will act on the prosthesis. Numerous attempts have been made to determine the forces acting on the hip-joint but the results are not reliable because the exact muscle action is unknown. Koch (1917) believed that the force acting on the hip-joint, when standing on one leg, is 1.6 times the body weight but Hackenbroch (1961) stated it is 4 times the body weight, while Williams and Lissner (1962) calculated it as 2.5 times the body weight. During walking, as a result of dynamic factors, the force is further increased and Hochman (1964), for instance, estimates it to be 6 times the body weight. Most calculations were made only in the frontal plane. The force acting on the hip-joint during walking and one-leg support is not purely vertical but it is believed to come from above in a downward, lateral, and ventral direction. Another drawback with calculations is that they have been made for a static position of the joint, e.g., with the subject standing on one leg or supine with the hip-joint flexed at a given angle.

Although under certain circumstances the force acting on the hip-joint can be determined to a fair level of accuracy by these methods, they have certain shortcomings. For instance, the calculations are usually made for only one plane. Thus, the results will be valid only in the cases in which the force acting in the third

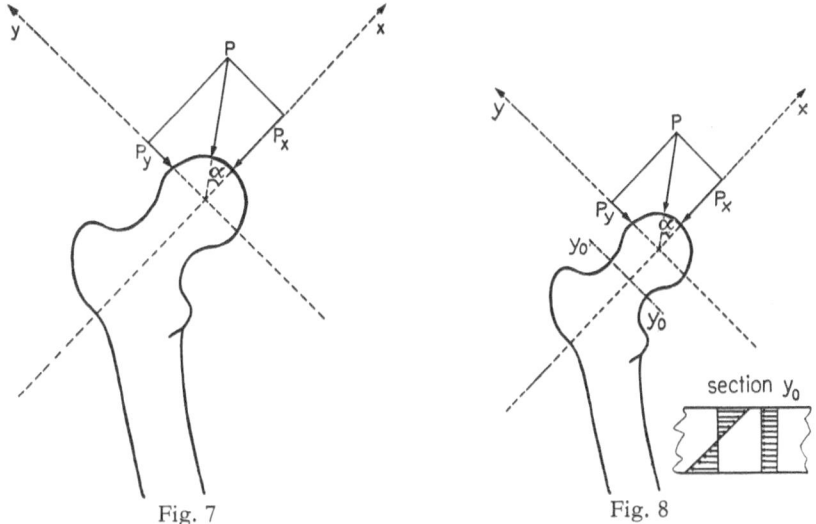

Fig. 7   Fig. 8

Fig. 7. Forces acting on the femoral head. In the frontal plane the force P can be resolved into two mutually perpendicular components P_x and P_y, parallel to a coordinate system fixed in relation to the prosthesis. A third component P_z goes into the plane of the figure perpendicular to P_x and P_y

Fig. 8. Forces acting on the femoral neck. The component P_x gives rise to compressive strains and P_y to both compressive and tensile strains in the neck

plane — the horizontal component — is negligible. If the values are to be accurate, it is necessary to calculate them for certain situations, e.g., with the subject standing on one leg, when the horizontal component is small and the force acting is large.

Though standing on one leg is a position in which the upper end of the femur is subjected to the greatest static force there are many others in which it is equally important to know the forces acting on the head of the femur. For instance, in flexion of the hip-joint with the leg straight the components P_x, P_y, and P_z should be nearly equal in magnitude. However, in this position application of the parallelogram of forces in one plane will not give very accurate values (Fig. 7).

Forces acting on the hip-joint are usually not static but, from the mechanical aspect, are a relatively slow dynamic development, the point at which the force acts on the head of the femur changing its position as the head moves in the joint. Since one component of force P acting on the hip-joint is the muscular force, the direction at which P acts on the head changes as the end of the femur moves in the acetabulum. In the different stages of walking — the swing phase and the stance phase — it is almost impossible to calculate the motion of the point of action of the force over the surface of the femoral head. In flexion and extension, or abduction and adduction, it is difficult to determine exactly the direction in which the muscles pull and the length of the moment arm, which introduces a factor of unreliability in the calculation of the forces acting in the hip-joint. If it were possible to measure these forces directly on the living subject, more reliable information would be obtained and the dynamic course of events could be recorded.

The force acting on the femoral head creates stresses in the neck of the bone (Fig. 8). In theory, therefore, it should be possible to determine the stress by means

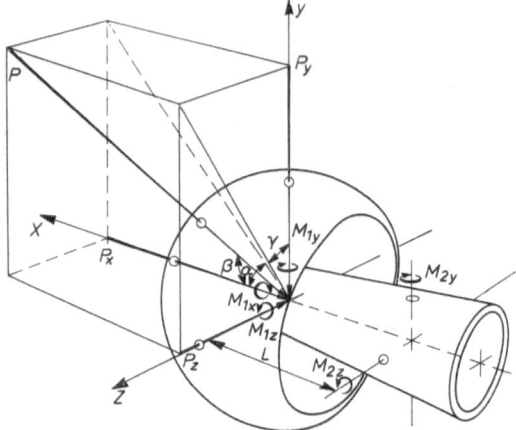

Fig. 9. Rectangular coordinate system of forces referred to prosthesis.
(See text for explanation)

of strain gauges applied to the neck of the femur, and from the recorded values to calculate the components of the force acting on the head. In practice, however, this is not feasible because problems arise in applying strain gauges under vital conditions, e.g., it is difficult to attach the gauges firmly to the moist and slightly greasy surface of the bone. The conducting material must be insulated against moisture and measures must be taken to avoid any toxic effect of the cement.

Since the physical properties of a bone differ not only from one person to another but also within different parts of the same bone, the measurements are not generally reliable. To obtain accurate measurements, a calibration must be performed by applying a known load, which is impossible in experiments in vivo.

Because metal is a highly suitable material as a support for a strain gauge and the head and neck of the femur can be replaced by a metal prosthesis, it was considered that the required forces might be determined by placing strain gauges in a prosthesis that could be inserted into the upper end of the femur.

The commercially available hip prostheses are not suitable for the author's purposes. Most of them have no neck which is indispensible because the essential feature of the author's method is that the forces acting on the hip-joint are determined from stresses set up in the neck, by the bending moment. If the measurements are to be accurate, the dimensions of the prosthesis must be exact and the strain gauges, which are not well tolerated by the tissues, must be applied so that they cannot come into contact with the body fluid. The best way of ensuring this is to place the gauges inside the prosthesis.

In the design of a prosthesis, replacing the upper end of the femur, account must be taken of the manner in which the signals are to be transmitted from the prostheses to the recording unit. Telemetry would have been the best and the most elegant method but was rejected on practical and financial grounds. The signals had, therefore, to be transmitted by wires in the conventional way.

Theory

In the design of a prosthesis of the type proposed with built-in strain gauges the following theoretical assumptions must be made. If P is the force acting on

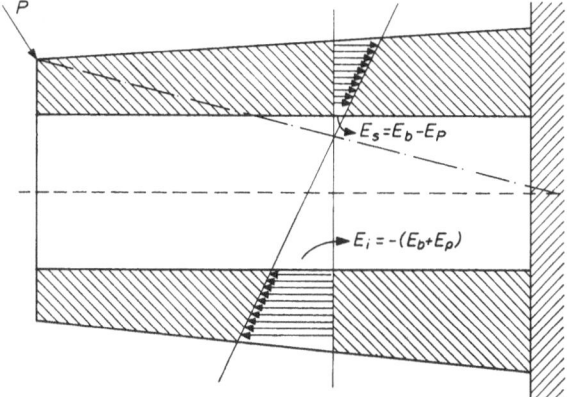

Fig. 10. Separation of forces from moments by electric strain gauges in a Wheatstone bridge. ε_s superior strains; ε_i inferior strains; ε_b tensile strains; ε_p compressive strains

the head of the prosthesis, it is necessary to find the magnitude and direction of this force (Fig. 9). The force can be resolved into three mutually perpendicular components — P_x, P_y, and P_z — parallel to a co-ordinate system that is fixed in relation to the prosthesis. The x-axis of the system is placed on the longitudinal axis of the neck of the femur, the y-axis perpendicular to it, and the z-axis perpendicular to both of them. The component P_x gives rise to a compressive force in the neck of the prosthesis, and the components P_y and P_z give rise to the moments M_x and M_y about the z and y-axes, respectively.

By placing the strain gauges at a known distance from the points of action of the component forces stresses, resulting from the moments of the forces, can be recorded and the components determined. The stresses can be separated (Fig. 10) by means of appropriate receptors and connections.

On the basis of the preceding theoretical assumptions a prosthesis was designed (Figs. 11, 12). The prosthesis, of a titanium stabilized stainless steel alloy, consists of ball-joint, neck, and marrow space anchorage of the Moore-type. The prosthesis sphere is $1^7/_8$ inches (47.2 mm) in diameter, the most common size used in arthroplastic operations on the hip-joint. The neck contains the strain gauges from which leads are brought out via a cemented flange sunk in the prosthesis which is sealed with resin. The leads, which are 22 mm in length, pass through a teflon collar threaded on the nipple and are joined at their free ends with a 16-pole female contact built into a hermetically sealed metal cylinder. The wires can be cut off near the prosthesis, when the measurements are completed, and removed. By calibrating the prosthesis, with and without the sphere, the distance between the tow sections of measurements could be determined.

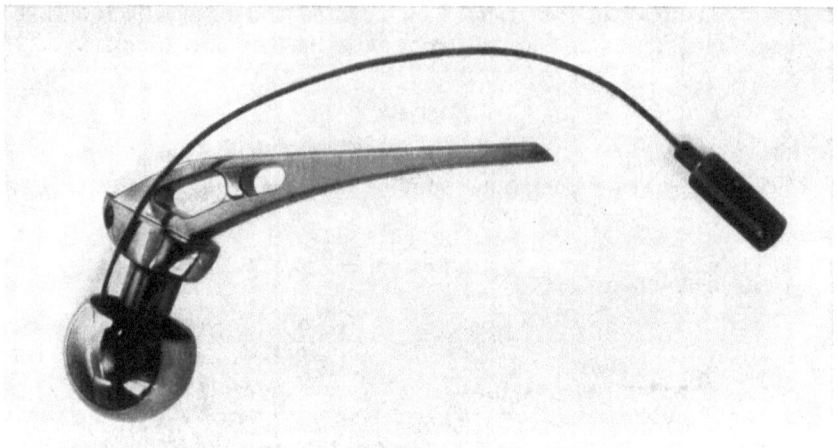

Fig. 11. The measuring prosthesis

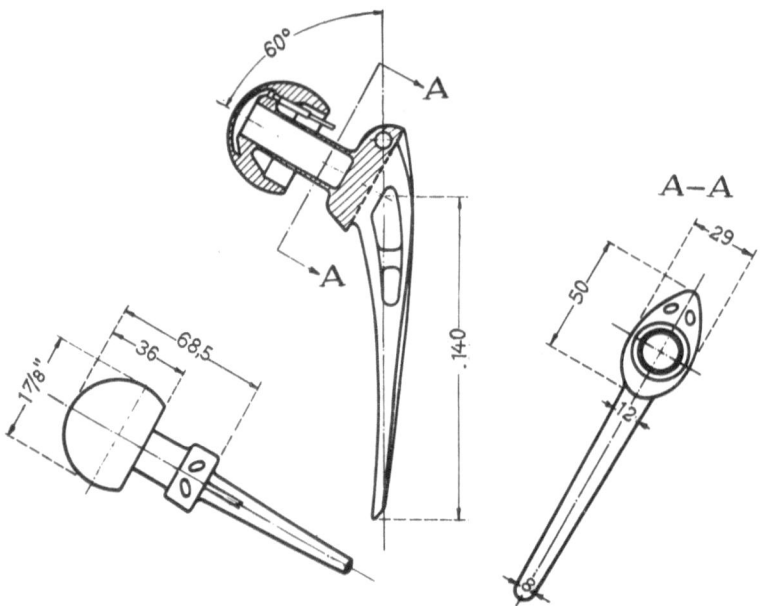

Fig. 12. Various views of the measuring prosthesis

Results

Two prostheses with built-in strain gauges have been placed into the hip-joints of two patients. Forces acting on the joint have been measured during different conditions. Both patients, a 52 year-old male (prosth. 1) and a 56 year-old female (prosth. 2) had a normal acetabulum. The man had a fracture through the femoral head, and the woman a medial fracture of the neck in a bad varus position. The author is fully aware that measuring data obtained from these prostheses cannot directly be transferred to normal conditions but they can give a quantitative understanding of the forces. The measurements were made half a year after the operation when the patients apparently walked normally without limp. The body weight of the man is 75 kp and that of the woman 44 kp.

Flexion

Fig. 13 shows the results for flexion in the supine position with the knee extended. The force is maximum when the leg has just started to move and, as seen from the figure, is greater than the body weight. The point of application of the force lies within a small area of the femoral head in the various experiments.

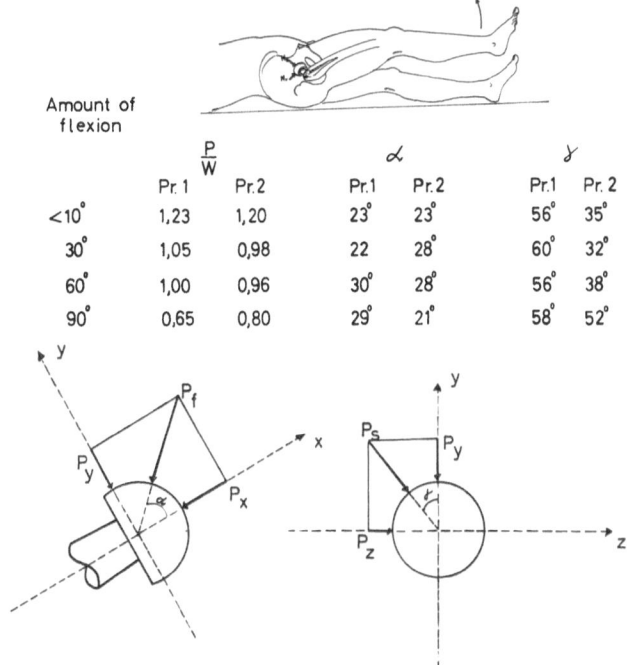

Amount of flexion	$\frac{P}{W}$		α		γ	
	Pr. 1	Pr. 2	Pr.1	Pr. 2	Pr.1	Pr. 2
<10°	1,23	1,20	23°	23°	56°	35°
30°	1,05	0,98	22	28°	60°	32°
60°	1,00	0,96	30°	28°	56°	38°
90°	0,65	0,80	29°	21°	58°	52°

Fig. 13. Forces acting on the hip-joint in flexion. P the resultant force; W body weight; P_f and P_s the projection of P in the frontal and sagittal plane respectively. For prosthesis 1 $W = 75$ kp. For prosthesis 2 $W = 44$ kp

Extension

The force acting on the hip-joint when the leg is extended is fairly high (Fig. 14). The point of application of the force is on the ventral side of the femoral head as in flexion in spite of the patient's being in the prone position.

Abduction

The values for abduction in the supine position are shown in Fig. 15. Abduction was performed on a flat surface with a low coefficient of friction. The force acting on the hip-joint is smaller than in flexion and extension but acts within the usual area of the femoral head.

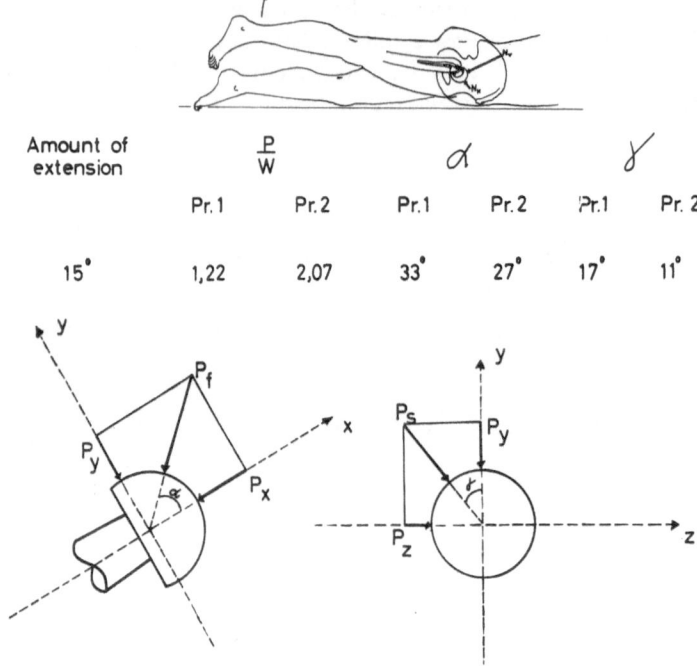

Amount of extension	$\frac{P}{W}$		α		γ	
	Pr. 1	Pr. 2	Pr. 1	Pr. 2	Pr. 1	Pr. 2
15°	1,22	2,07	33°	27°	17°	11°

Fig. 14. Forces acting on the hip-joint in extension. Symbols as in Fig. 13

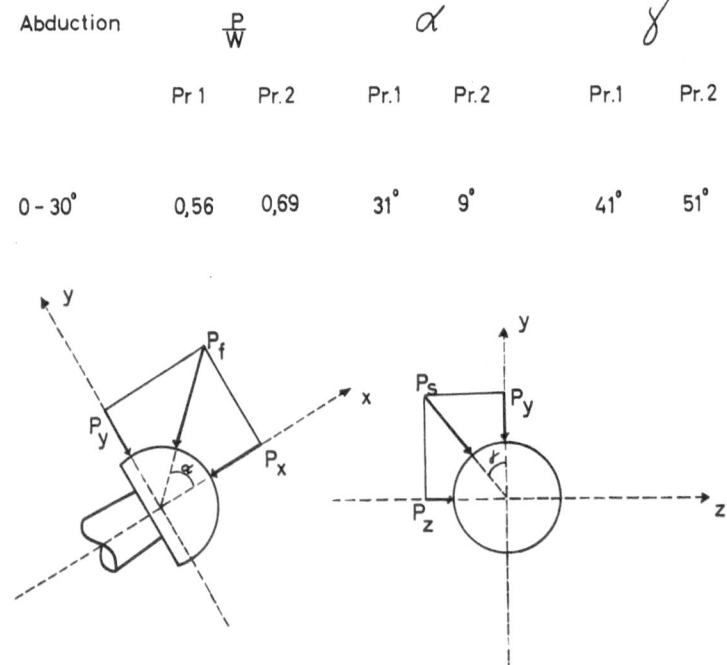

Abduction	$\frac{P}{W}$		α		γ	
	Pr 1	Pr. 2	Pr. 1	Pr. 2	Pr. 1	Pr. 2
0 - 30°	0,56	0,69	31°	9°	41°	51°

Fig. 15. Forces acting on the hip-joint in abduction. Symbols as in Fig. 13

Adduction	$\frac{P}{W}$		α		γ	
	Pr.1	Pr.2	Pr.1	Pr.2	Pr.1	Pr.2
$0-30°$	0,43	0,98	$24°$	$25°$	$45°$	$9°$

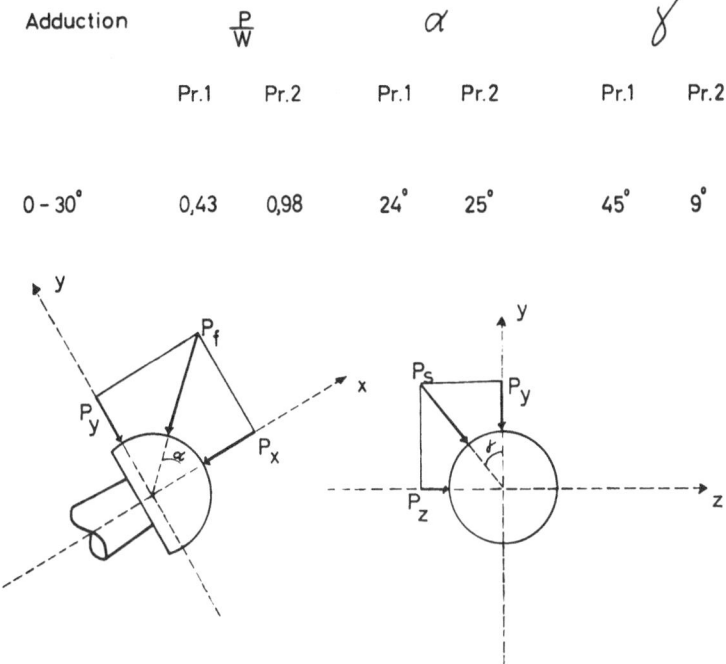

Fig. 16. Forces acting on the hip-joint in adduction. Symbols as in Fig. 13

Abduction opposite leg	$\frac{P}{W}$		α		γ	
	Pr.1	Pr.2	Pr.1	Pr.2	Pr.1	Pr.2
$0-30°$	0,25	0,19	$43°$	$22°$	$25°$	$44°$

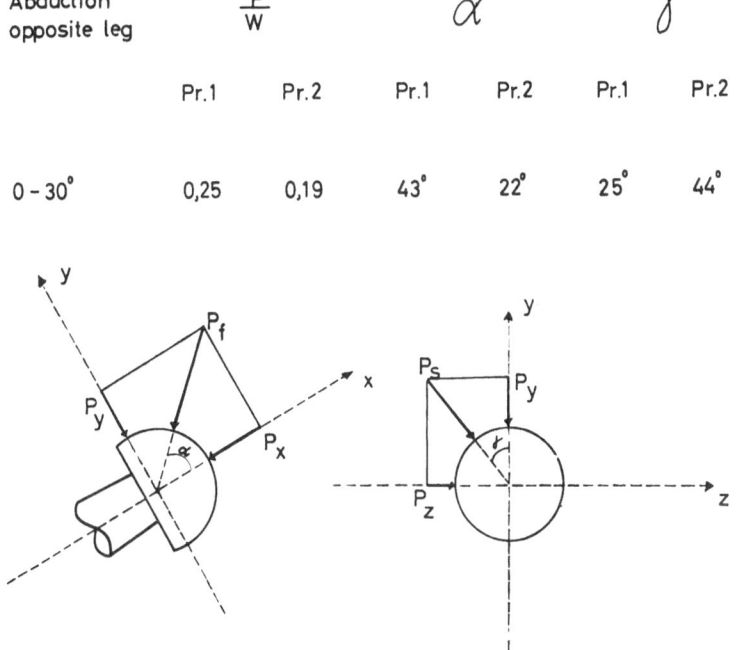

Fig. 17. Forces acting on the hip-joint in abduction of the opposite leg. Symbols as in Fig. 13

Adduction opposite leg	$\frac{P}{W}$		α		γ	
	Pr.1	Pr.2	Pr.1	Pr.2	Pr.1	Pr.2
0 – 30°	0,39	0,89	42°	33°	19°	7°

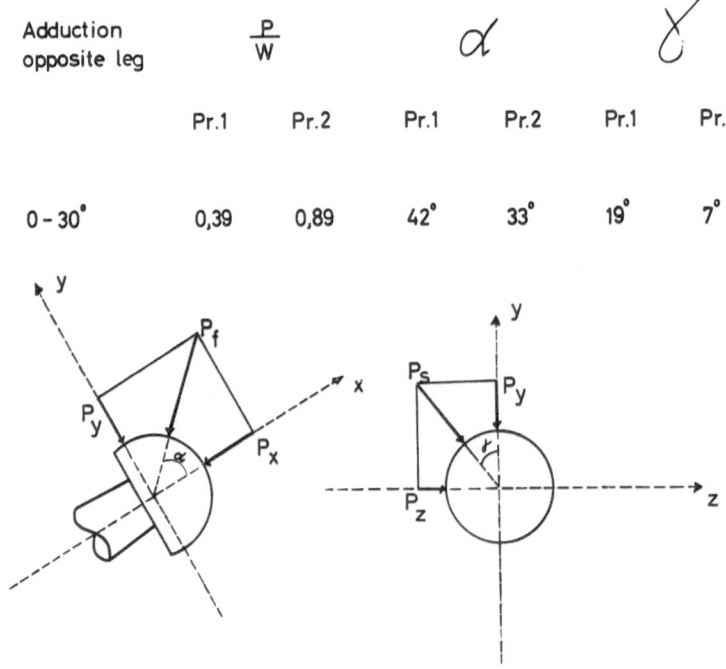

Fig. 18. Forces acting on the hip-joint in adduction of the opposite leg. Symbols as in Fig. 13

One leg support

	$\frac{P}{W}$		α		γ	
	Pr.1	Pr.2	Pr.1	Pr.2	Pr.1	Pr.2
	2,3	2,8	41°	33°	7°	10°

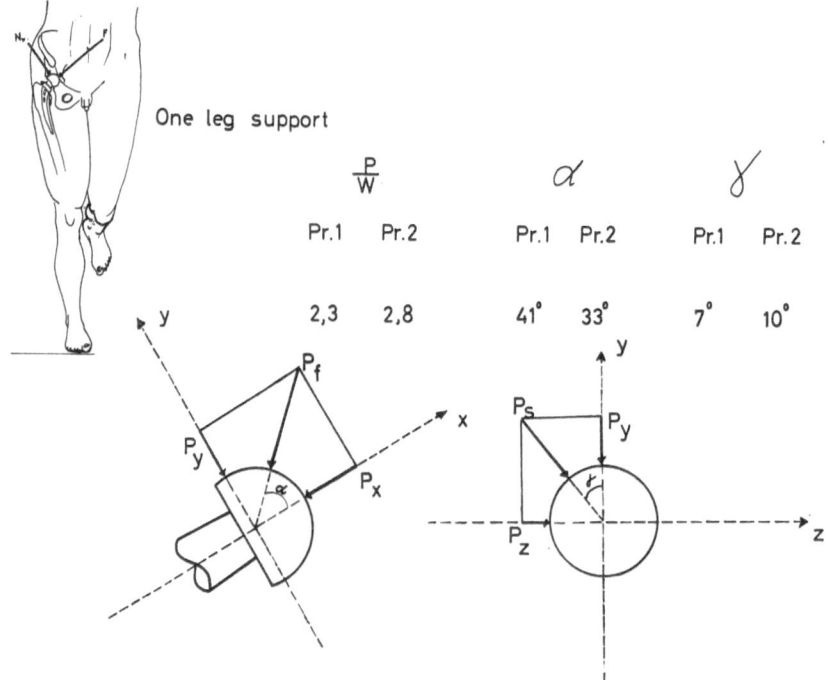

Fig. 19. Forces acting on the hip-joint when standing on one leg. Symbols as in Fig. 13

Adduction

The values for adduction are seen in Fig. 16. The difference in force between the two cases depends partly on the degree of flexion occurring in the second case.

Abduction, Opposite Leg

In abduction of the opposite leg with a patient in a supine position the force on the hip-joint is rather small (Fig. 17).

Adduction, Opposite Leg

When the opposite leg was adducted with the patient in a supine position a comparatively large force was recorded (Fig. 18).

Standing on One Leg

One leg support gives a large load on the hip-joint (Fig. 19). The force is about 2.6 times the body weight which corresponds well to that obtained on theoretical grounds. The horizontal force coming from the ventral side is extremely small and can be neglected.

Level Walking

The force acting on the femoral head during walking varies in magnitude and direction during a double step. In the support-phase stresses created in the hip-joint

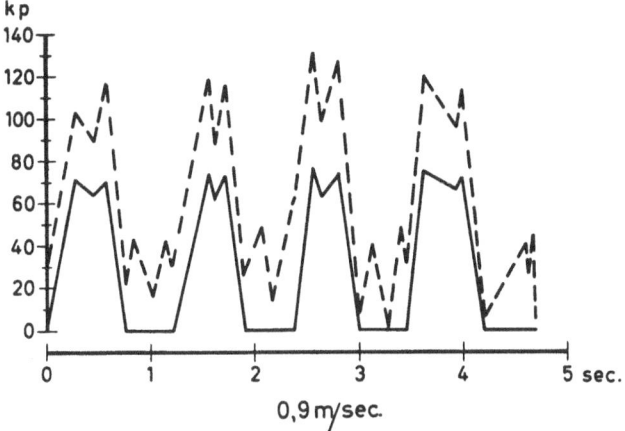

Fig. 20. Broken line: The force acting on the hip-joint during walking. Continuous line: The force between the foot and ground. Walking speed: 0.9 m/sec

are probably large, and because walking is one of the most common causes of great loads in this region, measurement of forces acting on the hip-joint during gait are important. In order to relate the force on the hip-joint to heel-strike, toe-off, support-phase, and swing-phase, two electronic force-plates, each 5 meters long, were constructed. By means of the force-plates the force between the feet and the ground as well as the force acting on the hip-joint were simultaneously determined.

During slow walking the force on the hip-joint was about 1.6 times the body weight, which is slightly more than that for flexion (Fig. 20). As can be seen from

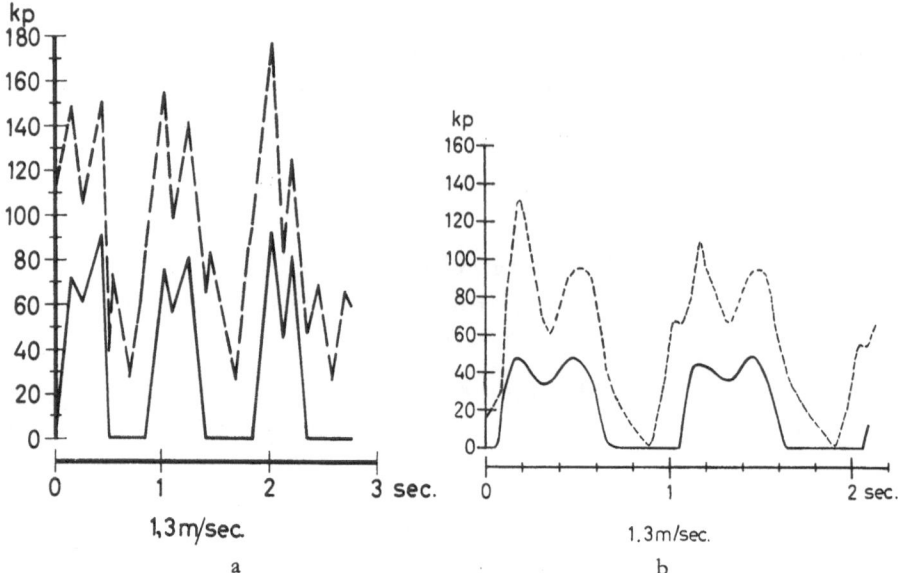

Fig. 21a and b. Broken line: The force acting on the hip-joint during walking. Continuous line: The force between the foot and the ground. Walking speed: 1.3 m/sec. a, Prosthesis 1. b, Prosthesis 2

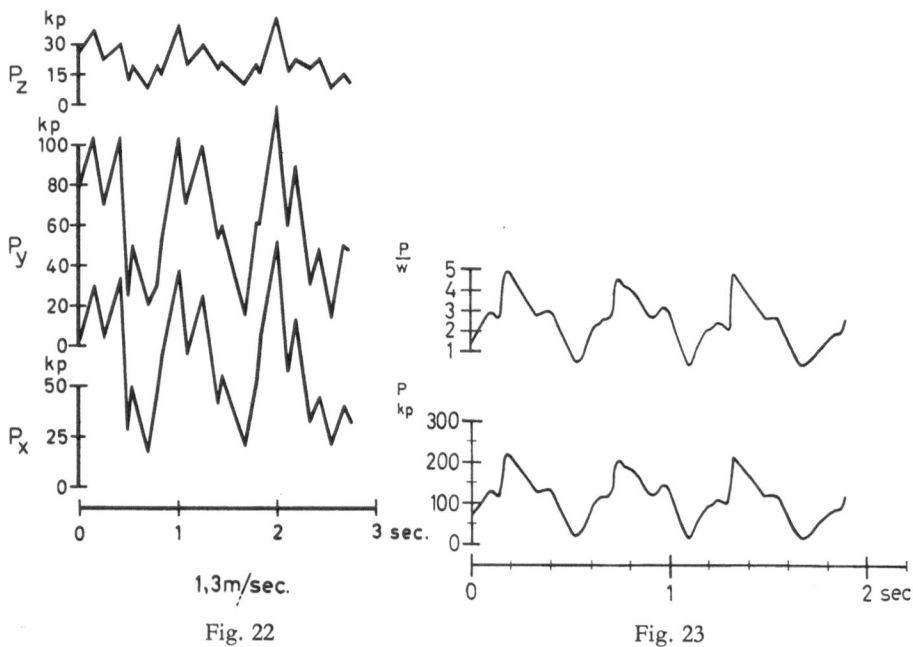

Fig. 22   Fig. 23

Fig. 22. The three components P_x, P_y, and P_z of the force P acting on the hip-joint during walking at a speed of 1.3 m/sec. Note: The component P_z always acts from the same direction

Fig. 23. Force acting on the hip-joint during running. Running speed: 2.6 m/sec

the figure the force during the swing-phase is not negligible. If the walking speed is increased the force on the hip-joint increases too, both in the swing-phase and in the support-phase (Figs. 21a, 21b, and 22). Very slow walking produced a force less than body weight. Using a stick decreased the force considerably below body weight.

Running

The force on the hip-joint of the woman was recorded during running. Forces up to 5 times the body weight were recorded during the support-phase and up

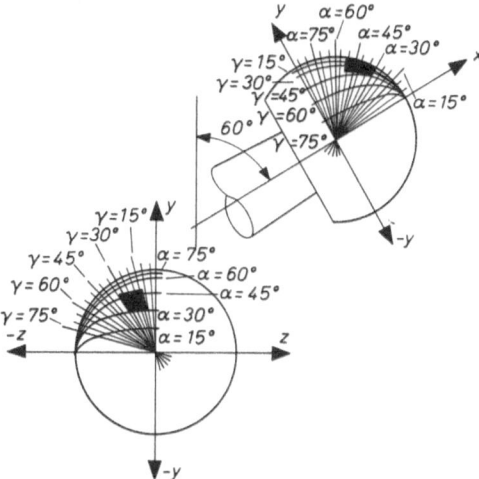

Fig. 24. The solid black area represents the range of variation in the points of application of the force P in all walking tests

to 3 times the body weight during the swing-phase (Fig. 23). During the course of a double step the force acted on a rather small area of the ventral, superior, and medial side of the femoral head (Fig. 24).

Walking Upstairs and Downstairs

Similar values were recorded in walking up and downstairs. The only important difference was that during the swing-phase in walking upstairs the force sometimes acted ventrally from beneath but its magnitude was very small.

Coefficient of Friction

The coefficient of friction was found to vary between 0.021—0.045. The values are surprisingly low as the coefficient of friction between two lubricated steel surfaces is 0.10 and between ice and steel in skating 0.03. The coefficient of friction in normal joints is about 0.009—0.015.

Summary

Intravital measurements of pressure within the hip-joint can be performed under certain circumstances. For this purpose two specially designed prostheses have been made and placed in two patients (a 52 year-old man and a 56 year-old woman). Half a year after the operation measurements were performed under different conditions.

During flexion in the supine position and extension in the prone position the force acting on the femoral head is higher than in very slow walking and walking with crutches. This means that a patient with a nailed hip-fracture can just as well be let out of bed as do exercises with the fractured leg. In addition, movements with the opposite leg give a rather high load on the femoral head.

In fast walking, the force on the femoral head increases considerably to about 2.8 times the body weight, a value corresponding closely to that for one leg support. During running the force can reach levels of more than 4 times the body weight. In the swing-phase of locomotion a fairly high force acts on the hip-joint. During ordinary walking the magnitude of the force slightly exceeds the body weight and during running is more than 3 times the body weight.

During all tests the horizontal component of the force acts on the head from a more or less ventral direction in accordance with the direction of the principal plane through the upper part of the femur. This may be one explanation for the inclination this plane forms with the longitudinal axis of the femoral shaft because the inclination gives the highest resistance against bending stresses. In the most distal part of the neck the lateral and the arcuate trabecular systems form a network in the shape of a T-beam, which suggests a construction of high resistances against bending.

References

BACHMAN, S.: The proximal end of the femur. Acta radiol. (Stockh.), Suppl. **146** (1957).

BILLING, L.: Roentgen examination of the proximal femur in children and adolescents. Acta radiol. (Stockh.), Suppl. **110** (1954).

BOURGERY: Traité complet de l'homme. Paris 1823.

HACKENBROCH, M.: Handbuch der Orthopädie, Bd. 4, S. 1—68. Stuttgart: G. Thieme 1961.

HOCKMAN, F. R.: Metals in the human body. Metals Review for August. 7—8 (1964).

KOCH, J. C.: The laws of bone architecture. Amer. J. Anat. **21**, 178—293 (1917).

MEYER, H. G.: Die Architectur der Spongiosa. Reichert u. Dubois-Reymond's Arch. 615—628 (1867).

PAUWELS, F.: Der Schenkelhalsbruch, ein mechanisches Problem. Stuttgart: Ferdinand Enke 1935.

ROUX, W.: Das Gesetz der Transformation der Knochen. Berl. klin. Wschr. **30**, 509—511, 533—535, 557—558 (1893).

— Gesammelte Abhandlungen über Entwicklungsmechanik der Organismen. Leipzig: Halle 1895.

RYDELL, N. W.: Forces acting on the femoral Head-Prosthesis. Göteborg 1966.

SMYTH, E. H. J.: The mechanical problem of the artificial hip. J. Bone Jt Surg. B **40**, 778—798 (1958).

WILLIAMS, M., and H. R. LISSNER: Biomechanics of human motion. Philadelphia: W. B. Saunders Co. 1962.

WOLFF, J.: Ueber die innere Architectur der Knochen und ihre Bedeutung für die Frage vom Knochenwachstum. Virchows Arch. path. Anat. **50**, 389—450 (1870).

— Langenbecks Arch. klin. Chir. **42**, 302—324 (1891).

— Das Gesetz der Transformation der Knochen. Berlin: August Hirschwald 1892.

— Die Lehre von der funktionellen Pathogenese der Deformitäten. Langenbecks Arch. klin. Chir. **53**, 830—905 (1896).

— Bemerkungen zu der vorstehenden Arbeit des Herrn Dr. BÄHR. Z. orthop. Chir. **5**, 60—65 (1897/98).

— Die Lehre von der funktionellen Knochengestalt. Virchows Arch. path. Anat. **155**, 256—315 (1899).

— Bemerkungen zur Demonstration von Röntgenbildern der Knochen-Architectur. Berl. klin. Wschr. **37**, 381—384, 414—417 (1900).

The Ergonomic Aspects of Articular Mechanics

M. A. MacConaill

For the purposes of this paper we can define ergonomics as the science and art of adapting machines, workrooms, and work-processes to workers. Too often in earlier days the workers had to adapt themselves to other things with a consequent unhappiness at best and permanent ill-health at worst. One of the earliest attempts to deal with this problem was made by JULES AMAR (1920), who was the Director of the Research Laboratory of Industrial Labour at the Conservatoire National des Arts et Métiers in Paris. His classic, *The Human Motor*, was published in 1914 and issued in English in 1920. Later, and with increasing momentum since World War II, ergonomic studies and their application have become worldwide and have embraced the psychological as well as the physical aspects of the affair. The present paper deals with a limited part of this physical aspect alone.

The Nature of Articular Ergonomics

All work, other than purely mental, depends upon movements at our joints. The consequences of this seeming platitude are not always apparent at first sight, as will be shown later. Moreover, most work involves simultaneous movements at a chain of joints *(chaîne articulaire, Gelenkkette)*, of which the individual joints are links *(chaînons)*. Even when the worker is seated at least the axial skeleton is also involved; and there is no need to emphasise the part played by the lower limbs in those who are not seated. It is, then, the movements at our joints that determine what is actually done by the worker in field or factory. The rehabilitation of the injured or the sick depends upon how the normal working of one or more joints can be restored, or, should this be impossible, how prosthetic use can be made of such joints as remain or can be replaced by mechanical substitutes. In the logical order, though not in the chronological, it is the movements of the joints together with the forces to be exerted and the speed of movement that determine what muscles are to be called into play. Hence the ergonomic aspects of articular mechanics are of fundamental importance.

In articular ergonomics there are two primary studies, the kinematic and the kinetic. The kinematic field has to do with articular movements and their consequences without respect to the forces involved. The kinetic field has to do with the forces involved in such movements, particularly with the muscular forces. Each of these fields of study has a subfield, namely, that dealing with the effect of those repetitive tasks which are so characteristic of modern industry, even in these days of increasing automation. All that can be done here is to outline some of the principles that should govern both the interpretation of work and motion studies and the consequent practical procedures designed because of them[1].

Ergonomic Articular Kinematics

Two phenomena of articular kinematics have a special bearing upon ergonomics. The first is the phenomenon of the *conjunct rotations* at certain joints, notably those of the shoulder and of the vertebral column. The second is the phenomenon of habitual movement and the consequent phenomenon of *congruent rotations*. These will be taken in sequence.

The Primary Postulate

A survey of the writings upon articular mechanics over the last 40 years or so indicates that it is necessary to lay down a primary postulate in terms of which both the kinematics and the kinetics of the subject should be studied. This postulate is: *all movements of bones take place at specific joints and are to be referred to these joints.* Although this postulate is in fact axiomatic failure to keep it in mind has introduced an unnecessary complication into many descriptions of otherwise excellent work and into the exposition of the conclusions drawn therefrom.

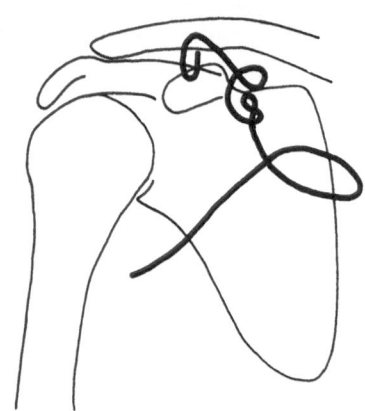

Fig. 1. The calculated path of the centre of rotation at the shoulder while the arm is being raised (after Provins)

For example, Provins (1965) shows a picture (Fig. 1) which purports to display the path of the instantaneous centres of rotation at the shoulder while the arm is being raised. This picture is typical of many that appear in motion-study publications. It is asserted here that it suffers from two major faults:

(1) The picture is two-dimensional whereas the bones at the shoulder (humerus, scapula and clavicle) have principal axes which, at the beginning of motion, are approximately at right angles to each other — the axis of the humerus being vertical, that of the relevant portion of the scapula being horizontal from before backwards, while that of the clavicle is horizontal but from the lateral to the medial side. This relationship cannot be adequately shown in a two-dimensional picture.

(2) The very complex course of the "centre of rotation" is no doubt valid for the mathematical construction derived from a study of successive photographs. But this pathway instead of helping us is actually a hindrance to our understanding, for it introduces a complexity where an intelligible simplicity is possible.

It is, in fact, not only simple but much more informative to remember that the raising of the arm at the shoulder is carried out by an articular chain: the humerus moves upon the scapula, the scapula moves upon the clavicle and the clavicle moves upon the manubrium sterni. The actual movement of the humerus is (in terms of angular velocities) the vectorial sum of these three movements. We have,

[1] Much of what follows is an amplification in detail of the account of articular motions given in *Synovial Joints* by Barnett et al. (1961).

in fact, a *gradient of angular velocities* as we pass from the sternoclavicular through the acromioclavicular to the glenohumeral joint, as was shown over twenty years ago (MacConaill, 1944) for the shoulder and still earlier for the wrist (MacConaill, 1941). These are two instances of the normal partition of movement between successive synovial joints, which is to be taken as the rule rather than the exception; for wherever there is a synovial joint there also is a potential motion, usually availed of as will be explained later. It may be remarked that the acromioclavicular joint tends to be forgotten by those studying the movements of the scapula, so that the so-called centre of rotation of the scapula has been assigned a variety of interesting but incorrect positions. It is in fact easier to visualise the separate though coordinated movements of these three joints than the celtic pattern traced by an invisible point which is itself the product of a series of mathematical operations often strange to those who have to apply their knowledge in practice. Moreover, when an assessment of dysfunction has to be made it must always be in terms of the motions at one or more of these three joints. In a word, anatomy cannot be wholly replaced by abstract conceptions.

Conjunct Rotation

When a bone is moved at a joint any point of it moves upon a curved surface in space, however small the curvature may be. This is a general law. Naturally, the curved motion of the bone upon its immediate articular companion may be nullified by the movement of this companion upon an oppositely curved surface in space so that the point we considered first may then trace a path upon a plane surface. We can, therefore, classify the movements of a bony point at one and the same joint into chords (geodesics upon the curved surfaces of motion) and arcs (non-geodesics upon the same surface)[1]. Chords correspond to straight lines in Euclidian geometry, arcs to all other lines in the same. Before discussing this fact further it is necessary to define the term *rotation* as it is used in the present context.

By "rotation" is meant the lateral (external) and medial (internal) rotation of medical anatomy; that is to say, movement of one articular surface around a normal to its fellow surface in the joint. For clarity this kind of rotation can be referred to as *spin* since one moving surface carries out a partial spin upon the other like a top upon a table. For the physicist and engineer such movements as flexion, abduction and their opposites are also rotations about some axis. Consequently some term expressing this fact is also desirable and any movement of a bone other than a spin will be called a *swing*. A swing may be defined, then, as any movement of the bone that does not involve a spin. There is no occasion in the present paper to consider those small movements of translation that also occur with swings, although they are important in connection with the theory of articular lubrication.

If a body moves around a straight line upon a flat surface it does so without any spin: this could be used as a kinematic definition of a straight line, more generally of a chord on any kind of surface. If we take any arc and join its extremities by a chord; and if we slide the body without any externally imposed spin upon this arc from one extremity to the other: then the body will be found to have undergone a spin as well as a translation between the two extremities of the arc. This, in turn, could be used for a kinematic definition of an arc, movements along chords

---

[1] This general use of the word *arc* is in accordance with its use in topology.

being pure translations. The spin that the body has undergone is one form of *conjunct rotation*, because it is necessarily conjoined with the movement along the arc. The amount of the conjunct rotation is the same whether the surface on which the arc is be plane or curved in any way: it is equal to the sum of the two angles subtended between the arc and the chord as in plane geometry. We could, of course, modify the amount of spin, making it greater or less, by a suitable, externally imposed spin. Such a spin would be called an *adjunct rotation*. If the amount of adjunct rotation were equal and opposite to that of the conjunct rotation, then the body would not only be in the same *position* but also in the same *posture* at the end of its journey along the arc as it was at the beginning. These simple and indeed familiar ideas are essential for the development of the present theme.

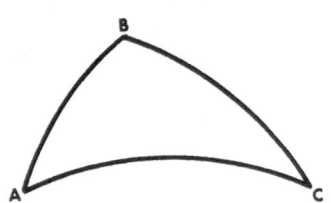

Fig. 2. A closed diadochal path in space of a point in a bone moving at a joint. Let the path be $A—B$, $B—C$, $C—A$. The portion $C—A$ is a "closing equipollent" (see text) to the other two parts if, and only if, the sum (S) of the three angles of the triangle is 180⁰. Otherwise the bone will undergo a conjunct rotation equal to (180°—S)

It follows from what has been said above that all paths described by a bone in space (at one and the same joint) are curved paths, in the ordinary sense of that term, whether they be chords or arcs. Consider now the following experiment with the upper limb, during which no pronation or supination whatever is allowed in the forearm; so that any movement of the arm will entail a corresponding movement of the whole upper limb. Move the arm in any direction, then move it a different direction and finally bring it back to its original position by a third movement. Any point on the limb will have then described a triangle in space, all of whose sides are curved. This triangle can be considered to lie upon some virtual surface for purposes of description. It can be shown (MacConaill, 1946) that two successive *(diadochal)* movements constituting two sides of any triangle are equivalent to a movement from the first point of the diadochal pathway to its last point if and only if the three internal angles of the triangle involved have a sum precisely equal to 180⁰. That is to say, in such a case, and only in such a case, the motion along the third side of the triangle is the *resultant* of the motion along the other two sides (Fig. 2)[1]. Any triangle having this sum of its three internal angles will be called a *Euclidian triangle*. A plane triangle is a Euclidian triangle drawn upon a plane surface or its equivalent (cylinder or cone). But we could draw Euclidian triangles upon a plane surface which are not of the type known as plane triangles; all their surfaces could be curved provided that they fulfill the condition regarding the angular sum.

Should a triangle not be Euclidian then the difference between its (internal) angular sum and 180⁰ will be called its *residual*. It is the residual that is of ergonomic importance. If a closed triangular pathway is followed by the limb then the said limb will undergo a conjunct rotation equal in amount to the residual. The sense of this rotation will depend upon two factors: first, whether the triangle be traversed

---

[1] This is the real basis of the triangle of displacements of ordinary mechanics, from which we get the triangles of acceleration and of forces used therein.

in a clockwise or an anticlockwise sense (viewed from some aspect); secondly, whether the residual be positive or negative in sign. If the triangle be traversed clockwise and the residual be positive then the conjunct rotation will be clockwise; but if the residual be negative and the triangle traversed clockwise then the conjunct rotation will be anticlockwise. The complete scheme is shown in Table 1.

When Table 1 is being read, it is to be noted *that the conjunct rotation is dependent upon the motions of the bone itself in space,* whether or not these are brought about by movements of a single joint. Thus the movement of the humerus to be described in the next paragraph is probably due to movements at three joints — the scapulo-humeral, the acromioclavicular and the sternoclavicular.

As an example of conjunct rotation due to diadochal movements we can take the following. Let the reader's right arm hang by his side so that his palm touches his thigh. Let him now swing the limb upwards through 90^0, then backwards through 90^0, and finally downwards through 90^0 until it hangs once more by his side. During these motions he must not pronate or supinate his forearm nor consciously medially or laterally rotate his arm. He will find that at the end of this closed triangle of movement his limb has been rotated laterally through 90^0. If he repeats the cycle, starting from the position of the limb in which it ended the first cycle, he will find that the limb has been rotated through 180^0 as a result of the two cycles. He will find, however, that he cannot perform a third cycle. The reason for this set of phenomena is as follows.

Table 1. *To show the relationship between the sense of a closed diadochal movement, i.e., cycle* $(\pm D)$, *the sign of the residual* $(\pm R)$ *of the corresponding closed figure and the sense of the consequent conjunct rotation* $(\pm C)$. *If* $+D$ *be taken as a clockwise movement then* $+C$ *and* $-C$ *indicate clockwise and anticlockwise conjunct rotations, respectively*

	$+R$	0	$-R$
$+D$	$+C$	0	$-C$
$-D$	$-C$	0	$+C$

During each cycle any point on his upper limb describes a triangle all of whose sides are curved and whose internal angles have a sum of 270^0. The residual of this triangle is therefore 90^0, the amount of the observed lateral rotation. From the reader's own point of view the triangle is described in a clockwise direction. From the same point of view lateral rotation is also a clockwise rotation. Hence both the magnitude and the sense of the conjunct rotation are as predicted. A third cycle cannot be performed because the two cycles cause the ligaments of the joints to become fully tightened in such a manner as to prevent any further movement, particularly of lateral rotation; this is ascertained by experiments upon osteoligamentous preparations. The joints are, in fact, in what is now known as the close-packed position (MacConaill, 1941) and the bones at these joints form a single mass virtually.

The term "sphere of movement of a joint" is well established in the literature. Although the surface intended by this phrase is not exactly a sphere, rather that of an ellipsoid, yet a triangular or any other pathway of the bone in space can be represented adequately enough upon a sphere. On such a sphere a chordal (i.e., geodesic) triangle has a residual that is always greater than 180^0, so that a motion traversing such a triangle will produce a clockwise or anticlockwise rotation in accordance with Table 1. The movements of flexion, backward rotation, and adduction of each cycle were swings, that is, movements along chords on the surface of motion; and the angle between each chord and its successor was 90⁰: hence, as stated above, the residual was 90⁰ per cycle.

If the reader now carries out a forward swing (flexion) of 90⁰ and then a backward swing through 90⁰, he will observe that his limb is in the same position and posture as if he had carried out a sideways swing (abduction) through 90⁰, together with a lateral rotation of the same amount. That is to say, the residual of the flexion and backward swing is a third swing between the first and last points of the previous diadochal movement together with a rotation, in this case lateral. Two inferences can be made from this fact:

(1.1) The lateral rotation occurs during the second limb of the diadochal movement, in this case during the backward swing.

(1.2) After the completion of backward swing the limb can be brought back to its original position *and posture* by some movement such that the resulting three-sided figure has a residual of zero (Fig. 2). This movement will *not* be along the shortest path from the end of the backward swing to the final position. It will be one of an indefinitely large number of *longer* paths, all of which will fulfill "the condition of zero residual", so to speak.

As already indicated, the experiment upon the upper limb described above is one exemplification of the working of the more general theorem (MacConaill, 1946, 1961, 1962—63). This theorem, already stated above, will now be restated in language more directly related to ergonomics.

We begin by drawing a distinction between the *resultant* of two (or more) successive movements and the *equipollent*. The two terms are usually considered equivalent in mechanics and are so, indeed, in mechanics involving movements in straight lines. But they are not equivalent in our present universe of discourse. We define them as follows:

(2.1) The *resultant* of a set of motions ($n-1$ motions) constituting ($n-1$) sides of an n-sided figure is the chord that completes the n-sided figure, together with a rotation whose magnitude is the sum of the internal angles of the n-sided figure diminished by $(n-2)\pi$. The chord is traversed from the first to the last points of the ($n-1$) system of paths; and the sense of the rotation is that of the direction in which the ($n-1$)-sided figure is traversed if the residual be positive, but is an opposite sense if the residual be negative.

(2.2) The *equipollent* of the ($n-1$) paths referred to in (2.1) is any path from the first to the last points of the ($n-1$)-system which completes an n-sided figure fulfilling the zero residual condition, i.e., such that the angular sum of the consequent n-sided figure is precisely $(n-2)\pi$.

The significance of the equipollent is this. If the moving bone be brought to a certain position by ($n-1$) successive movements then passage back to its starting position along an equipollent path will undo any conjunct rotation it has undergone, so that the limb is returned again both to its original position and its original posture.

The Ergonomic Significance of Conjunct Rotation

Space has been devoted to a discussion of conjunct rotation, leading to the concept of the equipollent, because it is a fundamental part of ergonomic kinematics. Practically every movement of a limb that is performing some task is a diadochal movement. This is because the limb has, in fact, to be raised to some extent before it grasps a tool, turns a wheel, pulls a lever, and so on. Hence the limb will either undergo a conjunct rotation or will require the exercise of muscles of lateral or medial rotation to prevent such a conjunct rotation. For example, if I sweep my hand, palm upwards, over a table and parallel to the surface of the table, I must first have raised my hand above the level of the table and then swept it either laterally or medially to begin with. In the (former) case my limb would undergo a lateral conjunct rotation, overcome by an unnoticed pronation of the forearm; in the later case what would be a medial conjunct rotation of my hand is avoided by a similarly unnoticed supination of my forearm. There is no need to develop this aspect of the matter further except to remark that it applies to all "working motions", including those of the eyes. In the case of the eyes it can be used to explain the presence of the superior and inferior oblique muscles (MacConaill, 1961). In these and similar instances we have an example of the working of synergisticmuscles, of a kind unsuspected until the notion of conjunct rotation had been fully developed.

Of more importance in the present context is what we may call *the closing equipollent of a work-cycle*. This is the pathway along which a limb (or part of it) is brought back to its original position at the last stage of a work-cycle, so that the succeeding cycle can begin again under precisely the same conditions as the first. As an example of such a work cycle you could take the placing of an object taken from a box and placed in another receptacle which is travelling on a conveyer belt. Here the movements will be the raising of the upper limb, a swing to one side to meet the oncoming container, the dropping of the object into the receptacle (which can be ignored), and the return of the hand and limb to the original position for the next cycle. The return motion will be an equipollent.

We have seen that a spherical (or quasi-spherical) triangle, whose sides are chords, must have a residual greater than zero. Hence the equipollent in a working cycle is not in general formed by a chord but by an "arc". That is to say, *the equipollent closing motion will not be along the shortest possible path from its beginning to its end*. In other words, what some of those engaged in work-study might regard as a fault to be corrected is, in fact, a necessity! Moreover, there is a very large number of possible equipollents whereas there is only one chordal (shortest) pattern. It may be reasonably assumed that a good worker will unconsciously choose that particular equipollent which he finds conducive to easy repetition of his work-cycle. There is no need to labour this point further in a paper of this kind. It is enough to show that there is a problem of correct interpretation of a given record of a work-cycle.

Congruent Rotations and Habitual Movements

Cognate with the problem discussed in the previous section is that posed by what we call congruent rotations, viz., of successive parts of a limb. It is closely connected with the idea of conjunct rotation.

If a man laterally rotates his humerus and also supinates his forearm then the movements at the shoulder and at the radiohumeral joints are called *congruent*. This is because supination is essentially a lateral rotation of the radius at the radiohumeral joint; so that both humerus and radius are carrying out rotations of a similar kind, even though the rates of humeral and radial lateral rotation may differ. From this example it will be clear that by a congruent rotation we mean the movement of two or more bones in series in the same sense, clockwise or anticlockwise. This matter has been discussed in some detail elsewhere with reference to its biological significance (MacConaill, 1950). We now consider its ergonomic implications.

At each joint there is a pair of movements that tend to be performed so constantly that they may be called *habitual movements:* and these are certainly characteristic of the joint. For example, we can take opposition and reposition of the thumb. It is these movements that we perform, to begin with, in our infancy, grasping or releasing some object. Only later do we learn to perform the somewhat artificial movements of radial and palmar deviation of the thumb and their opposites. Furthermore, the cycle consisting of (radial deviation+return) followed by (palmar deviation+return) if repeated rapidly soon tends to transform itself to a simple (opposition+return). Hence we can say that opposition and reposition are the characteristic and habitual movements of the thumb. As all know, in grasping some object, say upon a table, it is usually necessary to pronate the forearm, that is, rotate the radius medially. Now opposition of the thumb entails a conjunct medial rotation of the first metacarpal bone. Here, then, we have a very common example of a habitual motion which involves congruent rotations of a preceding bone (radius) and a succeeding bone (metacarpal). In this case the adjunct rotation of the radius is congruent with the conjunct rotation of the metacarpal in opposition able to be brought about by a radial swing of the thumb, followed by a palmar swing of the thumb at right angles of the first motion. Thus we could also say that, if the first carpometacarpal joint were one of three degrees of freedom then the medial rotation observed in opposition would be the rotation congruent with the two swings mentioned being carried out in the order given.

But we can go further. Let a man hang his upper limb loosely by his side and let his thumb be held loosely against the rest of his hand. Now let him pronate his hand quickly, but without looking at it. In most cases he will find that his thumb has swung into opposition as a result of this sudden medial rotation of his radius. That is to say, his thumb has carried out a rotation congruent with that of the radius. In terms of physics, the angular momentum of the rotating radius has been communicated through the wrist to the thumb which, being free to move, acquires an angular momentum in the same sense as that of the radius. For it to remain in its original position against the hand would require the expenditure of muscular energy, as can be verified by experiment; furthermore, if the thumb be put into opposition to begin with it will stay in opposition as a result of the radial movement already described. We are dealing here with the operation of a physical law of least expenditure of energy: it is, so to speak, easier for the thumb to rotate medially (which entails the opposition movement) than not to.

Similarly, it can be shown that the characteristic movements of the upper limb as a whole are an upward and outward swing at the shoulder, together with supination; and the reverse movement once the first has been accomplished. Here we have congruent lateral rotations of the humerus and radius in the "upswing" and the reverse congruent rotations in the "downswing". Again, the habitual, or certainly characteristic, movements of the free lower limb are an upward swing with lateral rotation and abduction with its opposite, a downward swing with medial rotation and adduction. Flexion of the femur at the hip when combined with or followed by abduction entails a lateral conjunct rotation, and conversely *mutatis mutandis* for the movements characteristic of the downswing. Thus these characteristic motions entail rotations that can be brought about as a result of a diadochal movement. The lateral or medial rotations involved are congruent with the resultant of the diadochal movements involved in upswing and downswing, as the case may be; and are of the same magnitude and sign as that of the equipollent of these diadochal movements.

It seems reasonable to assert that explicit notice should be taken of the habitual motions at joints and of the associated phenomenon of congruent rotation both in the design of machines and in the evaluation of work-studies based upon the use of existing machines. These habitual movements are clearly those of least constraint from the point of view of the worker. They are less likely to lead to fatigue because of this fact and should be respected as far as possible in the design not only of machines but also of what could be well called "working space". MORANT (1955) has drawn specific attention to this factor, saying that "the arrangement (spatial) should be such that all movements can be carried out without strain and with no difficult sequences". He was referring in particular to the design of aeroplane cockpits, which in these days could well include rocket cabins. But what he said obviously applies to all working spaces whatsoever, in the light of what has been said above. This point will be reexamined towards the end of this paper.

Ergonomic Aspects of Myokinetics

As its name implies, myokinetics has to do with the actual muscular forces used to bring about a movement at one or more joints. Space does not allow of more than a brief reference to those parts of the subject that are of ergonomic interest. Because of the radical change in our ideas regarding the actual employment of muscles on various circumstances, a change brought about by the increasing volume of verified electromyographic studies, much new thinking will have to be done upon this supremely important matter. There appears to be a growing agreement between the more modern theory of muscle action and experimental evidence. It seems safe to say that muscular action, in particular instances, is governed by certain *principles of minimal action*. The theoretical side has been treated of in some detail by MACCONAILL (1949, 1956, 1957, 1962/63, 1964, 1965). A large part of the experimental evidence will be found summarised in BASMAJIAN (1962); the aerlier work of himself and others will be found in this book. Nearly every number of the *Journal of Bone and Joint Surgery* carries fresh confirmatory evidence bearing on the same topic.

The minimal principles are as follows:

(3.1) No more total muscular force is used than is both necessary and sufficient for the task to be performed, whether this be one of supporting some weight or carrying out a movement, resistance to which may vary from zero upwards. This is the *Principle of Minimal Total Muscular Force*.

(3.2) No more spurt muscles are employed than are necessary and sufficient for the task to be performed. This is the *Principle of Minimal Spurt Force*. (See below.)

(3.3) No more shunt muscles are employed than are necessary and sufficient for the degree of speed at which a movement is performed. This is the *Principle of Minimal Shunt Force*. (See below.)

(3.4) Within one and the same muscle no more fibres are employed than are necessary and sufficient for the muscle to perform its allotted task. This is the

Principle of Minimal Fibre Number. It could also be referred to as the principle of minimal myoneural units (of SHERRINGTON), because the number of muscle fibres brought into play depends upon the number of neurones stimulated.

Theorems (3.2), (3.3) and (3.4) are really the constituent parts of (3.1). Numbers (3.2) and (3.3) may require some explanation for many readers.

By a *shunt muscle* is meant one which is normally brought into play only when motion at the joint on which it acts is rapid[1]. Its functional origin is near the joint on which it acts and its insertion is far from the joint on which it acts, so that the greater part of its contractile force is directed along the bone(s) over which it passes and in the direction of the joint at which it acts as a shunt muscle. It provides the necessary increase of centripetal force called for by rapid movement, as distinct from slow (MacConaill, 1949, 1957; Basmajian, 1959). A *spurt muscle* is defined as any muscle that is not a shunt muscle; it begins far from the joint on which it acts and is inserted near it. In slow motion it supplies enough centripetal force along the moving bone on which it acts to satisfy Newton's law of curvilinear motion. At the elbow joint the brachialis and the brachioradialis muscles are very typical examples of spurt and shunt muscles, respectively.

Proceeding from the notion of spurt and shunt muscles, we have those of spurt and shunt *forces*, i.e., the rectangular components of a given muscular force that are at right angles to the bone(s) acted upon and along the said bone(s), respectively. Two muscles capable of the same kind of action may exercise very different amounts of shunt force. For example, the pronator quadratus and the pronator teres can both rotate the radius medially. But the quadratus is almost horizontal, whereas the teres is much more vertical. The teres will then exercise a greater shunt force than the quadratus. Accordingly, it is found by electromyography that the quadratus alone is used in slow pronation, the teres being also brought into operation in rapid pronation and pronation against marked resistance; similarly, the supinator (brevis) is alone used in slow supination, the biceps being also used in rapid supination against marked resistance (DE SOUSA et al., 1958; TRAVILL and BASMAJIAN, 1961).

The examples just given illustrate both the laws of minimal spurt and minimal shunt action. The pronator quadratus and supinator are typical spurt muscles for the movements they produce; and they alone are used unless there be need to supplement them because of increased speed of movement, weight to be moved, or other increased resistance. Similarly the brachialis is the normal flexor of the elbow; supplemented by the brachioradialis as a shunt muscle in rapid movement, and by the biceps when there is supination against resistance, together with flexion of the elbow. It will be observed from these examples that there always appears to be a muscle which is normally given more work todo than any others anatomically capable of performing the same action. This appears to be a general law, a fact strikingly exemplified by the relatively small supraspinatus at the shoulder. As BASMAJIAN and BAZANT (1959) have shown, this muscle is that normally employed for supporting the pendent

---

[1] In Spanish the terms "spurt" and "shunt" (muscle) have recently been translated as *músculo de impulso* and *músculo desvío*, respectively. In French the corresponding renderings have been *muscle moteur* and *muscle statique*. The latter translation is not a very happy one, since shunt muscles are primarily used during rapid movements and not for the maintenance of a given posture unless the load carried be very great indeed.

limb even when it is carrying a considerable weight, although it may be helped in many individuals by the most posterior fibres of the deltoid: this fact has a manifest bearing upon the frequency of painful dysfunction of this muscle.

We can say, therefore, that for every joint there are "slave muscles" *(musculi serviles)* which have to work always while others can, and do very frequently, take their ease. The number of such servile muscles will depend, of course, upon the degree of freedom of motion at the joint concerned. There will be flexor-extensor, abductor-adductor and medial and lateral rotator pairs of such muscles in the case of such joints as the shoulder and the hip. This fact would appear to have importance in ergonomic theory and practice, for it would explain the occurrence of fatigue in workers performing what would otherwise be thought to be not very strenuous, repetitive tasks. It is clear that here there is a field where electromyographic and workshop studies will require further development and correlation. It is in this field that articular kinetics has its special place.

The Possible Need for Varied Tasks

In these days a very high proportion of those working in factories and workshops are engaged in performing the same kind of task all through the day. There are undoubtedly intervals for the taking of tea or coffee, with the accompanying rest implied thereby, in the more humane countries. But this rest is largely passive and is not enough from the point of view of the arthrologist. The proper nourishment of *all* of the joints requires movement and this may be lacking when the employment is of a sedentary kind, even although it involves movements of several joints of a given limb. The cyclic nature of such movements necessarily restricts the amount of movement called for and not only the joints of the whole body but many muscles of the limbs involved may be insufficiently exercised in such a way as to promote, amongst other things, a thorough exercise of, say, the blood circulatory system. Notwithstanding some recent arguments to the contrary, it would appear to be reasonably established that during movement of a joint there is a corresponding internal movement of the synovial fluid which promotes the maintenance of the articular cartilages in a healthy state. For all these reasons it would seem a good thing to give each worker, especially a sedentary worker, alternative tasks which would involve movements of his lower as well as of his upper limbs and vigorous movements of both these and of his trunk.

In those countries in which the working week has been reduced to 5 days there is a theoretical possibility that this exercise could be carried out on the two remaining days of the 7. But the increasing substitution of passive for active participation in games and athletic contests, together with the increasing use of automobiles for transport to and from work, as well as for leisure purposes, makes this substitute somewhat doubtful in modern urban communities, certainly in Western Europe and North America. We have not a few joints but many; and all require to be exercised if the whole man is to be healthy. Today as always we anatomists are concerned with whole and living men rather than only with dead parts.

Summary

(1) The paper deals with two ergonomic aspects of articular mechanics, the kinematic and the kinetic.

(2) The kinematic aspect arises from the phenomenon of Conjunct Rotation and the cognate phenomenon of Congruent Rotations along a serial sequence of and the joints *(Gelenkkette)*.

(3) Arising from the foregoing there are certain preferred movements of limbs and their parts which should be taken account of by ergonomists.

(4) The ergonomic aspect of articular kinetics is governed by a three-fold principle of minimal action: minimal spurt action, minimal shunt action, and a minimal employment of muscle fibres within the muscles brought into play.

(5) Many modern factory and workshop programmes can lead to insufficient exercise of the whole articular and muscular system by those engaged at repetitive tasks. In addition, all joints require constant exercise if they are to be sufficiently nourished and kept healthy, which makes it desirable for alternative tasks to be provided for those in certain types of industry for the sake of the general health of every worker.

References

Amar, J.: The human motor. London: George Routledge & Sons 1920.

Barnett, C. H., D. V. Davies, and M. A. MacConaill: Synovial joints, their structure and mechanics. London: Longmans 1961.

Basmajian, J. V.: "Spurt" and "shunt" muscles: an electromyographic confirmation. J. Anat. (Lond.) **93**, 551—553 (1959).

— Muscles alive. Baltimore: Williams & Wilkins Co. 1962.

—, and F. G. Bazant: Factors preventing downward dislocation of the adducted shoulder joint. J. Bone Jt Surg. A **41**, 1182—1186 (1959).

MacConaill, M. A.: The mechanical anatomy of the carpus and its bearings on some surgical problems. J. Anat. (Lond.) **75**, 166—175 (1941).

— The mechanical anatomy of the acromioclavicular joint in man. Proc. roy. Irish Acad. B **50**, 159—166 (1944).

— Studies in the mechanics of synovial joints. II. Displacements on articular surfaces and the significance of saddle joints. Irish J. med. Sci., Ser. VI, Jul. 223—235 (1946).

— The movements of bones and joints. II. Function of the musculature. J. Bone Jt Surg. B **31**, 100—104 (1949).

— Rotary movements and functional décalage. Brit. J. phys. Med. **13**, 50—56 (1950).

— Studies in the mechanics of synovial joints. V. The statics of single joints. Irish J. med. Sci. Aug. 353—364 (1956).

— Studies in the mechanics of synovial joints. VI. Motion at one joint. Irish J. med. Sci. Mar. 99—113 (1957).

— Mechanical anatomy of motion and posture. In: Therapeutic Exercise, p. 44—87 [S. Licht (ed.)], 2. ed. New Haven: Elizabeth Licht 1961.

— The mechanics of locomotion and posture. In: Annual volume of physiology and experimental medical sciences, p. 19—84. Calcutta: Physiol. Soc. Ind. and Soc. Exp. Med. Sci. Ind. 1962/63.

— Étude anatomique des mouvements et des postures. In: La thérapeutique par le mouvement. Paris: Cercle d'Études Kinésithérapiques 1964.

— Anatomía mecánica de la movilidad y de la postura. In: Terapéutica por el Ejercicio. Barcelona: Salvat Editores, S.A. 1965.

Morant, G. M.: Body measurements in relation to work spaces in aircraft. In: Anthropometry and human engineering. London: Butterworth Sci. Publ. 1955.

Provins, K. A.: Ergonomics for industry. 7. Men, machines and controls. London: Dept. Sc. Ind. Research 1965.

Sousa, O. M. de, W. R. de Morais e E. D. de F. Ferraz: Estudo eletromiográfico de alguns músculos do antebraço durante a pronação. Rev. Hosp. Clín. **13**, 346—354 (1958).

Travill, A., and J. V. Basmajian: Electromyography of the supinators of the forearm. Anat. Rec. **139**, 557—560 (1961).

A Longitudinal Vital Staining Method
for the Study of Apposition in Bone *

M. J. Baer and J. L. Ackerman

Although vital staining with alizarin compounds has been used to study the nature of the growth of bone for more than two centuries, the method has undergone remarkably little change. Only two stages of technological development are clearly evident since John Belchier (1736a; 1736b) confirmed the fact that the madder plant stained the bones of living animals a red color.

Beginning with the inquiries of Duhamel (1739) and culminating in the work of Payton (1932, 1933) and Brash (1934), the madder plant was fed to animals of different ages and then withheld for a determined period of time prior to sacrifice. As a result, the bones of the experimental animals contained a band of red stain succeeded by a band of unstained white bone formed after the time that the dye was being absorbed.

The second stage in the development of the vital staining technique arose from the artificial production of alizarin, the essential dye of the madder plant. According to Cameron (1930), alizarin was synthesized and introduced as a substitute for madder for dyeing cloth as early as 1869. Thus the possibility existed for the administration of the dye to animals by injection rather than by the prolonged method of oral feeding. However, its actual use as a vital stain appears to have been deferred until Gottlieb (1914) tested the rate of uptake of alizarin by intravenous and subcutaneous injections in dogs. Finally, Schour (1936) and his colleagues (1938, 1941) showed that the intraperitoneal administration of alizarin Red S (sodium sulphalizarate) gave superior results when compared with subcutaneous, intramuscular or intravenous injections. The reliability of intraperitoneal administration and the commercial availability of alizarin Red S certified for staining bone generated interest in the use of the method for the analysis of skull growth. A variety of mammalian species have been studied including the rat (Massler and Schour, 1951), the monkey (Moore, 1949; Craven, 1956), the pig (Mednick and Washburn, 1956), and the rabbit (Hoyte, 1961).

The feeding of madder obviously precluded the systematic study of the sites and patterns of bony deposition during the early postnatal period. Thus, the intraperitoneal injection of alizarin represented an important technical advance because the vital stain could be administered prior to weaning without the necessity for

* This investigation was supported in part by General Research Fund Grant FR 05320-01 and 05320-04, National Institutes of Health, and in part by Public Health Service Research Grant HD 01845-01, from the National Institute of Child Health and Human Development, Bethesda, Md.

forced feeding. No one, to the present writers' knowledge, has attempted to inject an extract of the madder plant itself.

The appeal of the vital staining technique lies in the possibility of depicting the pattern of postnatal bony deposition for an extended period of time in one animal. The fact that repeated feeding of madder to the same animal would permit the longitudinal study of bone growth was apparent to DUHAMEL in 1742. By splitting the long bones of a madder-fed animal, he observed alternate layers of red and white bone indicating growth in circumference by surface apposition.

The present technique can be used effectively in the longitudinal study of relatively simple appositional systems such as the calcification of dentin (SCHOUR, 1936) or the formation of the outer circumferential lamellae in a long bone (WEINMANN and SICHER, 1947). In these instances, the adjacent lines of stain are easily distinguishable and stand out sharply reminiscent of the annual rings in the trunk of a tree. For the detailed analysis of growth changes in the craniofacial skeleton, however, the repeated injection of alizarin Red S is much less effective. The extensive remodeling of morphologically complex areas of the skull makes it difficult to discriminate between the results of successive injections because the lines of stain appear to mingle. Investigators employing the technique of vital staining have, therefore, resorted to a modified longitudinal method in which a number of animals of varying ages are given a single injection and sacrificed at different intervals (HOYTE, 1960).

BRASH (1939) recognized that the longitudinal use of vital staining would be enhanced considerably if a different stain exhibiting a distinctive color reaction could be used at each injection. To this end, he tested ten derivatives of hydroxyanthraquinone and reported that five of these gave positive results, "each producing a characteristic colour". BRASH's findings have never been applied in the study of skeletal growth.

Recently, CLEALL et al. (1964) have devised a longitudinal vital staining technique which employs a variety of bone marking agents. By administering alizarin Red S, Terramycin, calcium[45], and trypan blue, singly and in different combinations, to growing rats, these investigators produced ten distinctive bone markers. They illustrate the application of four of these indicators used in sequence to show apposition in a simple growth site. As CLEALL and his co-workers have pointed out, however, incorporation of these types of marking agents entails the use of light microscopy, ultraviolet light microscopy and autoradiography, in the analysis of the data.

This paper reports a longitudinal method designed to extend the effectiveness of the present vital staining technique with alizarin Red S while maintaining its inherent simplicity. To achieve this objective, an alizarin dye was sought which would give a contrasting color reaction to alizarin Red S when injected into living animals and which could be analyzed in the same way. At the suggestion of Dr. ROBERT M. STEPHAN, National Institute of Dental Research, Bethesda, Maryland, the dye, acid alizarin blue BB was tested and was found to meet these requirements.

Methods and Materials

Alternate injections of alizarin Red S and of acid alizarin blue BB (C. I. No. 58610) were administered intraperitoneally to each animal in a series of white rats (Sprague-Dawley). A 2% solution of the alizarin Red S and a 1.5% solution (a saturated aqueous solution) of the acid alizarin blue BB were used in a dosage of 1 cc per 100 gms of body weight. As many as four injections were given to the same animal at fixed time intervals. The injection schedule was programmed to reveal the pattern of bony apposition during the first 130 days of post-natal life.

Following sacrifice of the animals, the unfixed heads were grossly defleshed by hand picking. Final maceration of the soft tissue was accomplished by the process of enzymatic digestion using crude papain. This consisted simply of placing the head in an 8 ounce jar of water containing approximately a half teaspoon of papain and heating at 45⁰ C for 20—24 hours. The heads of rats under 20 days of age were cleaned completely by hand, to avoid maceration of the extensive sutural connective tissue and disarticulation of the bones. Cleaned skulls were stored in 95% alcohol.

The crania and disarticulated mandibles were photographed, while immersed in alcohol, using Agfachrome daylight film and electronic flash illumination. A number of other films available commercially were also tested but failed to photograph satisfactorily the blue dye deposited in the bones of even heavily stained specimens.

Crania were prepared for serial, undecalcified sectioning in the following manner. The specimen was cut in two parasagittally with a fine dental disc. The larger portion containing the midline structures was dehydrated in absolute alcohol for 24 hours prior to embedding in Ward's Bioplastic. In order to ensure infiltration of the plastic into the smaller interstices of the bones, the specimen was immersed for 24 hours in a solution of Bioplastic diluted with an equal part of acetone. The specimen was then placed in pure Bioplastic without catalyst for an additional 24 hours. At each stage the specimen was vacuumed intermittently to remove air bubbles. The specimen was then transferred to a rectangular mold in which a plastic base had already been formed. The embedding layer of fresh Bioplastic containing catalyst was poured and the plastic medium was cured for 12 hours at 37⁰ C. Final curing was achieved by heating at 70⁰ C for 14 hours.

Due to the fragility of the skull in the young rat, the crania of animals of less than 20 days of age at time of sacrifice were embedded whole.

The plastic embedded skulls were sectioned with a Brownell thin-sectioning machine. The unpolished sections were mounted on microscope slides with Per-mount and cover-slipped. Sections of 150—200 μ were found to be of optimal thickness for microscopic study using incident (reflected) lighting. Thinner sections, naturally, did not evidence as great an intensity of stain. Since the distribution of the two dyes at the surface of each section was the concern of this phase of the study, rather than the cellular content of the bones, thinner sections offered no real advantage. Photomicrographs of each section were also made with Agfachrome film and electronic flash illumination.

Observations

Figs. 1 and 2 (Plate 1) show the cranial vaults of white rats injected at one day of age with alizarin Red S and acid alizarin blue BB, respectively, and sacrificed at ten days of age. The acid alizarin blue BB results in a distinctive blue-orchid

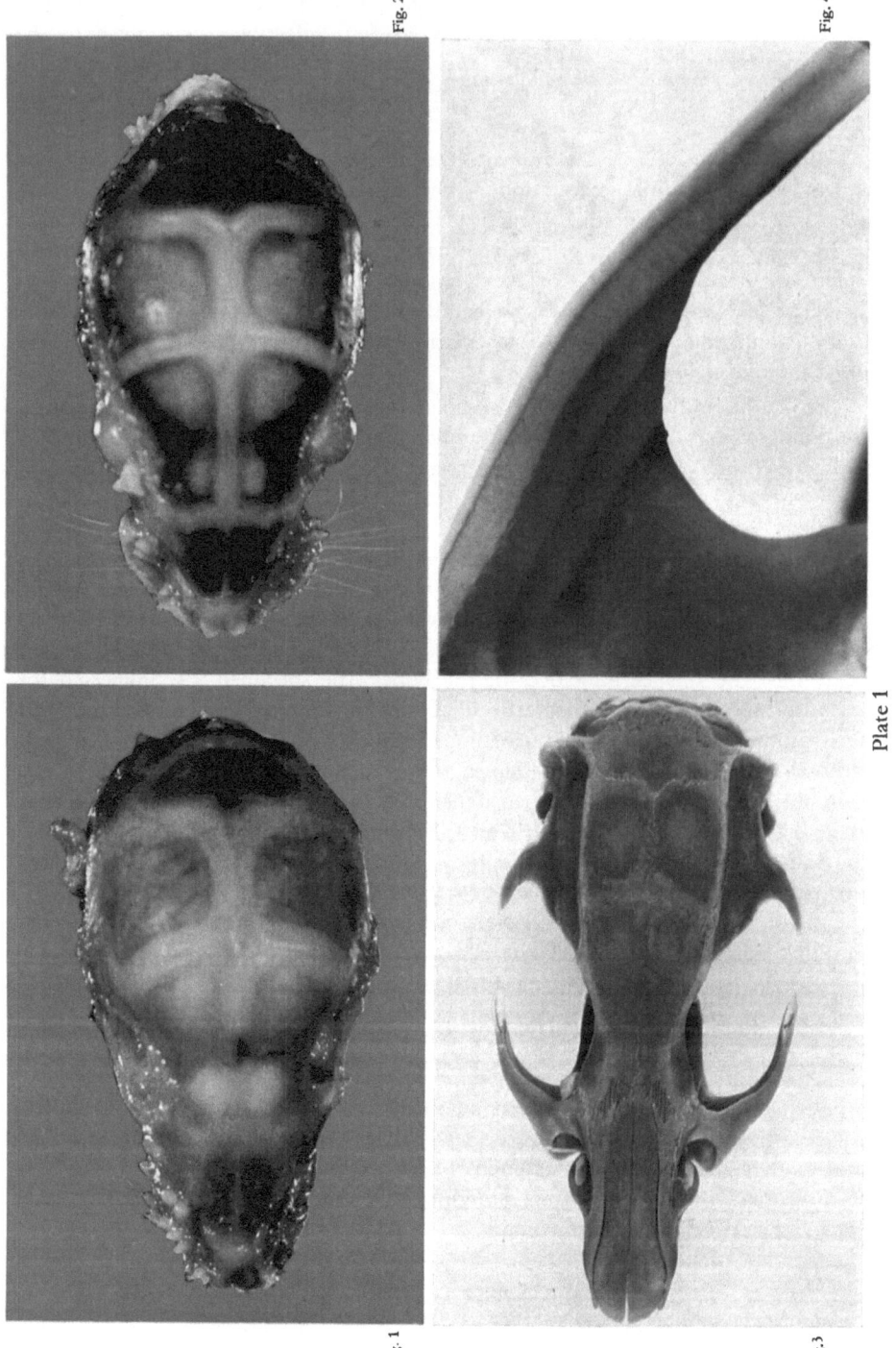

Fig. 2

Fig. 4

Plate 1

Fig. 1

Fig. 3

coloration in sharp contrast to the red-pink staining of the alizarin Red S. The stains evidence the same pattern of distribution in the cranial bones. Both vaults exhibit bands of unstained white bone representing the incremental growth which occurred at the sutural margins of each center of ossification subsequent to the period of absorption of the stains.

Upon decalcification, both dyes were removed from the bones of even heavily stained specimens. Although the precise mechanism responsible for the incorporation of alizarin Red S into growing bone is still unresolved, it is generally accepted that this dye is associated with the inorganic phase of bone (CAMERON, 1930; SCHOUR et al., 1941). This viewpoint has recently received experimental support from MYERS' (1964) demonstration that alizarin Red S will form a chelate ring with hydroxyapatite *in vitro*. It is reasonable to assume that the acid alizarin blue BB reflects the deposition of mineral in the growing bone in a comparable fashion to the alizarin Red S.

Fig. 3 (Plate 1) illustrates the dorsal aspect of the skull of a rat which received a series of four injections (red — 30 days, blue — 47 days, red — 68 days, blue — 103 days) and was sacrificed at 126 days of age. Superficially, the cranial vault exhibits a purple coloration, due to the superimposition of the differently stained apposed layers of cortical bone and the translucency of the bone in the wet specimen.

Detailed examination of specific anatomic structures and undecalcified sections, however, indicates clearly that the dye introduced at each injection is sharply segregated and confined to the bone formed at the time of absorption of the stain. This fact is easily documented by examining the ventral aspect of the zygomatic arch in the same animal.

In Fig. 4 (Plate 1) the appositional growth history of the arch from 30 to 126 days is revealed at a glance. Further, it is possible to discern the precise location of the lateral margin of the arch at five chronological ages (the ages at the time of the four injections and the age at sacrifice). In the present specimen, the red stain representing the lateral margin at 30 days and the blue stain depicting the lateral margin at 47 days, now terminate abruptly at the anterior border of the temporal fossa; the red stain representing the third injection at 68 days now borders the medial aspect of the free portion of the arch. Lastly, the injection of blue dye at 103 days is banded by unstained bone resulting from the final 23 days of growth prior to sacrifice of the animal. It is evident that the zygomatic arch increases in transverse diameter by a process of surface apposition along the superolateral margin.

Fig. 1. Dorsal view of the skull of a rat injected at one day of age with alizarin Red S and sacrificed at 10 days of age. White bone at the sutural margins represents growth which took place subsequent to the absorption of the dye

Fig. 2. Dorsal view of the skull of a rat injected at one day of age with acid alizarin blue BB and sacrificed at 10 days of age

Fig. 3. Dorsal view of the skull of a 126-day old rat which received four injections of dye (red — 30 days, blue — 47 days, red — 68 days, and blue — 103 days). The cranial vault exhibits a purple coloration, due to the superimposition of the differently stained layers of cortical bone and the translucency of the bone in the wet specimen

Fig. 4. Enlargement of the ventral aspect of the zygomatic arch of the animal shown in Fig. 3. The position of the lateral margin of the arch at each injection is indicated by the alternately colored bands of stain

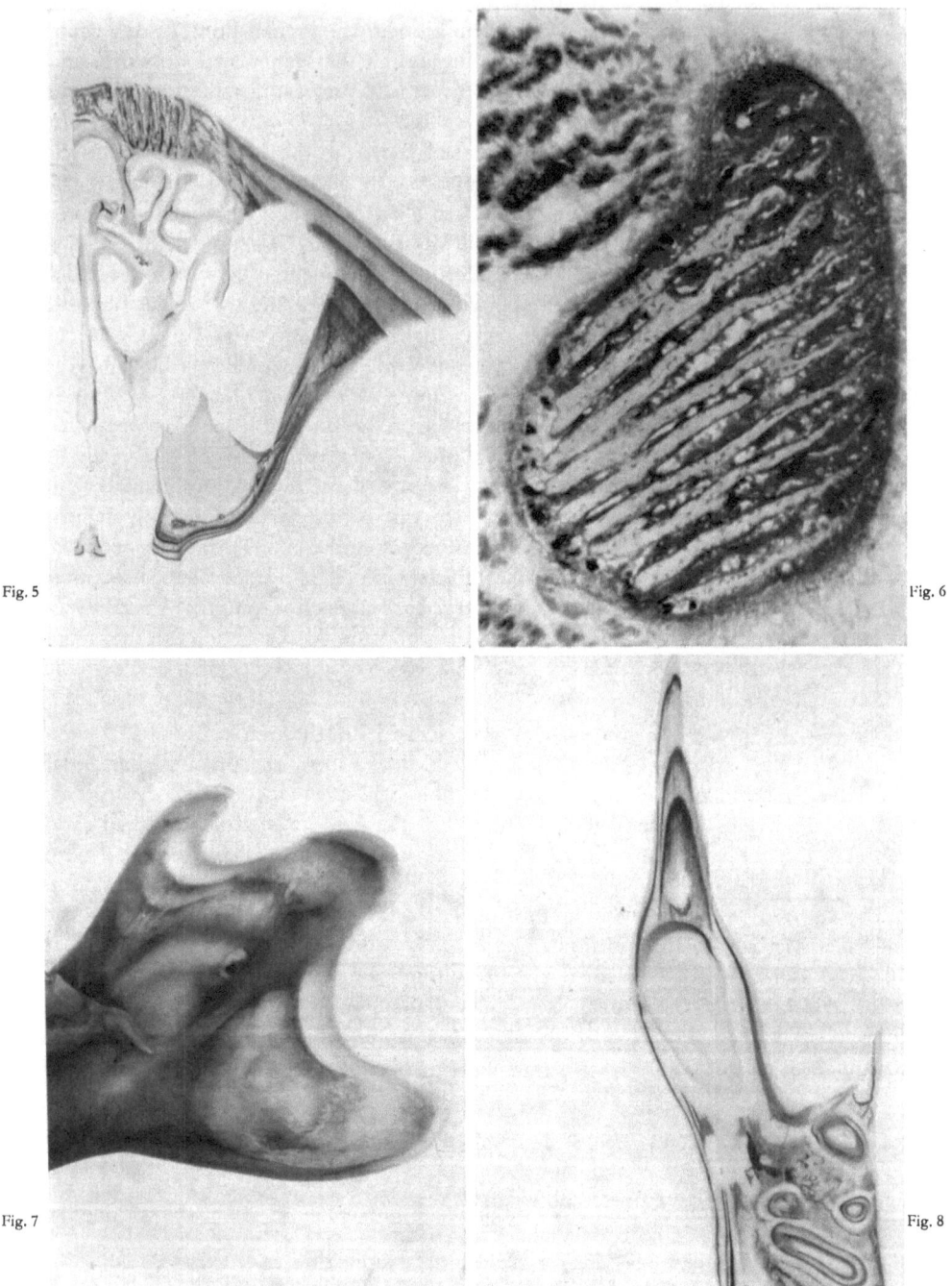

Fig. 5

Fig. 6

Fig. 7

Fig. 8

Plate 2

A detailed picture of the specific sites and pattern of apposition in the arch and associated structures may be obtained by studying serial sections of a longitudinally stained animal. Fig. 5 (Plate 2) shows such a section taken through the maxillary root of the arch in another rat in which the color sequence of injection was reversed (blue — 30, red — 47, blue — 68, red — 103, and sacrificed at 132 days). Each band of stain clearly demarcates the position of the superolateral margin at the time of injection. At the palatal surface, apposition lateral to the anterior palatine foramen has increased the vertical height of the maxilla. Superior to the turbinates, one may observe the complex pattern of staining of the interdigitating spicules in the nasofrontal suture.

BAER (1954) and DIXON and HOYTE (1963) used a graded series of animals each with a single injection of alizarin Red S to construct a composite picture of the mode of growth of the zygomatic arch. These writers concluded that the arch grew by a process of apposition on the superolateral margin and resorption from the inferomedial margin, resulting in the continuous replacement of bone tissue and the structure of the arch itself. Erosion of the medial aspect of the arch thus increases the transverse diameter of the temporal fossa. The longitudinally stained rat shown here in Fig. 5 incorporates, in a single specimen, the data to be obtained from a whole series of animals injected at different ages. In the specific case of the rat arch the pattern of staining and the obliteration of the lines of stain representing the lateral margin at earlier chronological ages provide indirect evidence indicating resorption of the medial border. For direct corroboration, however, it is necessary to demonstrate the concurrent resorptive activity per se.

Fig. 6 (Plate 2) shows a cross section through the free portion of the zygomatic arch in a 10 day old rat stained for succinic dehydrogenase activity and counterstained for alkaline phosphatase activity using the method described by FULLMER et al. (1964). The osteoclasts, selectively stained a deep brown, stand out in bold relief against the red background. This section reveals a continuous row of osteoclasts concentrated along the inferomedial aspect of the arch. The histochemical technique thus confirms the inference drawn from the pattern of deposition of the alizarin dyes, concerning the major site of resorption in the zygomatic arch of the rat.

In the mandible, the longitudinal staining technique provides a graphic picture of the sites, modes and directions of growth responsible for dimensional change.

Fig. 5. Coronal section through the maxillary root of the zygomatic arch of a 132 - day old rat which received four injections of dye (blue — 30 days, red — 47 days, blue — 68 days, and red — 103 days). The bands of stain show the mode of growth of the arch and the palate

Fig. 6. Coronal section through the free portion of the zygomatic arch of a 10 - day old rat stained for succinic dehydrogenase activity and counterstained for alkaline phosphatase activity. This section shows a continuous row of osteoclasts concentrated along the inferomedial aspect of the arch, stained a deep brown

Fig. 7. Medial aspect of the ramus of the mandible of the 126 - day old rat shown in Fig. 3. The alternately colored bands of stain reveal the positions of the tip of the coronoid process, the mandibular notch, the condyle, and the angular process, at each age of injection

Fig. 8. Horizontal section through the posterior half of the left mandible in a 132 - day old rat taken at the level of the molar roots and the superior part of the angular process. This specimen received four injections of dye (blue — 30 days, red — 47 days, blue — 62 days, and red — 103 days). Note pattern of staining in the ramus

Fig. 7 (Plate 2) illustrates the medial aspect of the ramus in a specimen which received four injections as follows: red — 30, blue — 47, red — 68, blue — 103,
and sacrificed at 126 days. It is possible to discern the entire ramal outline (the tip
of the coronoid process, the mandibular notch, the position of the condyle, and
the angular process) at several ages. While the mandibular notch is particularly
well defined for all four injections, the outlines of the condyle and the angular
process are only discernable for the last three injections. Posterior growth of the
mandibular foramen and the internal pterygoid ridge and the remodeling of the
sites of attachment of the masseter and internal pterygoid muscles have obliterated
the stain marking the position of the condyle and the angular process at 30 days.
The fact that the whole posterior border of the ramus comprises the major
site of growth responsible for lengthening the mandible is indicated by this
specimen.

While examination of the gross alizarinated specimen affords a general picture
of the growth of the ramus, sectioning is necessary to fully appreciate the complex
pattern of bony deposition in this structure. Fig. 8 (Plate 2) shows a horizontal
section through the posterior part of the angular process. This specimen also
received four injections of dye at nearly the same ages as the animal illustrated
in Fig. 7, although the color sequence was reversed (blue — 30, red — 47, blue — 62,
red — 103, and sacrificed at 132 days). The orientation of the slide places the posterior
border of the ramus at the top of the picture and the lateral surface of the mandible
to the right. Within the ramus, the white bone laid down subsequent to the last
injection, as well as the deposition of dye resulting from the last three injections
(red-blue-red), are easily discerned. The dye of the first injection, marking the position
of the posterior border of the ramus at 30 days of age, has been obliterated due
to the growth of the root of the incisor. Thus, only the faintest suggestion of blue
dye may be observed adjacent to the enamel layer.

The most interesting information revealed by this section, however, relates
to the pattern of remodeling activity on the medial and lateral aspects of the angular
process. While the band of red stain of the last injection extends along the entire
lateral aspect of the ramus and the body of the mandible, it terminates abruptly
on the medial aspect near the posterior border. The bands of stain representing
the two preceding injections, on the other hand, are intact and define both surfaces
at earlier ages. Examination of the section under high magnification, reveals that
the medial surface of the ramus posterior to the blue stain of the third injection
is scalloped and irregular, characteristic of a bony surface undergoing resorption.
The small area of white bone over the blue stain of the third injection indicates
accretion on the surface of the ridge representing the highest point of insertion
of the internal pterygoid muscle. Sections taken below this ridge through the
fuller part of the angular process show scalloping over the entire medial surface.

At birth, the angular process of the rat mandible consists of a fairly straight
slab of bone. During its postnatal growth it is subjected to extensive remodeling
so that at adulthood the lateral surface is convex and the medial surface is concave.
This change in form is brought about by the concurrent processes of apposition
on the lateral surface and resorption of the medial surface. The alteration in form
of this part of the rat mandible is strikingly similar to the remodeling of the basal
part of the ramus in the human jaw, as described by Enlow and Harris (1964).

Discussion

The student of skeletal growth has available to him a wide variety of methods and techniques from which to choose a suitable research procedure. In the area of vital staining and bone marking, techniques involving the use of radioisotopes, tetracyclines, and lead acetate, as well as alizarin Red S, have each received extensive treatment in the literature. Selection of a method to implement a particular research goal depends upon two criteria: the type of information which the method provides and the facility with which the techniques can be applied. If the techniques are too involved and cumbersome, they tend to discourage the investigator whose primary concern is the nature of bone growth and morphological change rather than the exploitation of a novel technique. It is necessary, therefore, to examine critically the attributes of the methodology described in this paper so that its usefulness may be assessed.

The primary value of multiple vital staining stems from the fact that it makes possible the longitudinal study of skeletal growth. The individual specimen contains a visually observable record of its growth covering a considerable period of its postnatal life. The actual changes in position of important anatomic landmarks can be observed directly in a single specimen. Consequently, the investigator is freed from the necessity of raising and injecting an extensive series of animals. Further, the problems of constructing a composite picture of growth from segmented data are largely eliminated.

Conversely, the multiple staining procedure shares the limitations inherent in the longitudinal method. An animal may die short of its programmed time of sacrifice after the investigator has expended considerable time and expense in its maintenance. Until the animal has been sacrificed and the bones exposed, the investigator cannot know for a certainty that the stain introduced at each injection was actually deposited in the growing bone. In the present study, animals occasionally evidenced only three bands of stain although they had received four injections. In some instances a blue band was missing, while in others the red stain was absent. Since there was no apparent pattern in the failure of deposition of the dye, the physiologic state of the animal at the time of each injection appears to be the significant factor which must be considered. The attrition of subjects, then, which plagues even the simplest of longitudinal studies must be anticipated in the planning of an experimental project using the method here described.

The major problems encountered at the technical level concerned the use of acid alizarin blue BB. This dye has a very low solubility in water (1.5%). Much of the material introduced at each injection remained localized in the peritoneal cavity in the form of a congealed mass. Also, it was found that not all of the dye lots of acid alizarin blue BB were effective in staining growing bone, even if they were obtained from the same manufacturer. Further work is necessary, therefore, to improve the solubility of this material and to determine the precise character of its staining properties.

In the opinion of the writers, however, the limitations in the method are offset by the simplicity of the technical procedures and equipment necessary for the analysis of the data. The distribution of both the red and the blue stains in the bone can be studied in the gross specimen with a dissecting microscope, while the detailed analysis of the sectioned material only requires the additional use of a light microscope with an incident light attachment. In practice, the analysis of the sectioned material obtained from one half of each skull can be made with the intact half at hand for reference in orienting the slide and identifying various anatomical relationships.

In addition to the rat, the multiple staining method was tested on rabbits, sheep, and pigs and found equally effective in demonstrating the pattern of bony deposition in these mammalian groups.

In each of these groups, the stain was retained in its original site of deposition throughout the experimental life of the animal, unless the site was subjected to subsequent resorptive activity. This suggests that metabolic turnover of calcium is not a factor which will remove the stain from its original site thereby destroying the longitudinal record of stain deposition.

Some investigators have rejected the vital staining technique with alizarin Red S on the grounds that the toxicity of the dye inhibits the normal growth of the animal (see discussions of this subject by Hoyte, 1960 and Dixon and Hoyte, 1963). The same argument could be raised against the use of alizarin blue BB since this dye has been shown to be toxic at high dosages and to have a suppressing effect on growth in weight (Cleall et al., 1964). The important point at issue here, however, is not the *rate of growth* but the *pattern of growth*. The vital staining technique reveals the manner in which bone is laid down and the manner in which change in form of osteological elements is brought about; it demonstrates the units of growth, the sites of growth, the direction of growth, and the relative duration of growth at different sites. Determination of the pattern of growth provides a meaningful basis for assessing the rate of skeletal growth in control animals free from the inhibiting effects of the dyes.

Some investigators have also criticized the vital staining method on the grounds that it does not provide direct evidence of the sites of bone resorption or of its extent. In the present study, it was found that the extent of resorptive activity could be clearly inferred in certain areas of the skull, such as the zygomatic arch or the palate, due to the pattern resulting from the interruption of the lines of stain. The extent of resorptive activity was much more difficult to infer, however, in other areas, particularly in the cranial vault of the rat, and especially, before the cortical plates became well defined. The importance of the resorptive mechanism in changing the shape of skeletal elements (Barnicot, 1947; Bateman, 1954; Enlow, 1963) stresses the need for its direct determination. The writers have found the techniques described by Barnicot (1947) and by Fullmer et al. (1964) for demonstrating osteoclasts to be complementary to the multiple vital staining method in the study of the concurrent processes of apposition and resorption.

Summary

This paper reports a two-color vital staining method designed to reveal the longitudinal pattern of apposition in growing bones.

Alternate injections of alizarin Red S and acid alizarin blue BB were administered intraperitoneally to the same animal at fixed time intervals. This procedure was tested on growing rats, rabbits, sheep, and pigs. Upon sacrifice, skulls representing all four mammalian groups showed bands of bone alternately stained red and blue indicating that the stain was deposited in the sites of growth active at the time of each injection. In a single rat mandible, for example, it was possible to discern the entire ramal outline (the tip of the coronoid process, the mandibular notch, the position of the condyle, and the angular process) at four different ages prior to sacrifice. Thin sections of undecalcified rat skulls exhibited sharply demarcated lines of stained bone in the cortical plates of the crania, as well as the mandibles, making possible a detailed analysis of the appositional growth history of discrete osteological elements.

It was concluded that the use of dyes of different colors facilitates the longitudinal study of the intricate pattern of bone deposition in the mammalian skull.

Acknowledgement

The writers are indebted to Dr. HAROLD M. FULLMER, National Institute of Dental Research, Bethesda, Md., for the histochemical material demonstrating osteoclasts.

References

BAER, M. J.: Patterns of growth of the skull as revealed by vital staining. Hum. Biol. **26**, 80—126 (1954).

BARNICOT, N. A.: The supravital staining of osteoclasts with neutral red: their distribution on the parietal bone of normal growing mice, and a comparison with the mutants grey-lethal and hydrocephalus-3. Proc. roy. Soc. B **134**, 467—485 (1947).

BATEMAN, N.: Bone growth: a study of the grey-lethal and microphthalmic mutants of the mouse. J. Anat. (Lond.) **88**, 212—262 (1954).

BELCHIER, J.: An account of the bones of animals being changed to a red colour by aliment only. Phil. Trans. **39**, 286—288 (1736a).

— A further account of the animals being made red by aliment only. Phil. Trans. **39**, 299—300 (1736b).

BRASH, J. C.: Some problems in the growth and developmental mechanics of bone. Edinb. med. J. **41**, 305—387 (1934).

— Vital staining of bone with hydroxyanthraquinone derivatives. J. Anat. (Lond.) **74**, 141 (1939).

CAMERON, G. R.: The staining of calcium. J. Path. Bact. **33**, 929—955 (1930).

CLEALL, J. F., R. E. PERKINS, and J. E. GILDA: Bone marking agents for the longitudinal study of growth in animals. Arch. oral Biol. **9**, 627—646 (1964).

CRAVEN, A. H.: Growth in width of the head of the macaca rhesus monkey as revealed by vital staining. Amer. J. Orthodont. **42**, 341—362 (1956).

DIXON, A. D., and D. A. N. HOYTE: A comparison of autoradiographic and alizarin techniques in the study of bone growth. Anat. Rec. **145**, 101—114 (1963).

DUHAMEL, H. L.: Sur une racine qui a la faculté de teindre en rouge les os des animaux vivants. Hist. Acad. roy. Sci. (Paris) 1—13 (1739).

— Sur le developpement et lacrue des os des animaux. Hist. Acad. roy. Sci. (Paris) 354—370 (1742).

ENLOW, D. H.: Principles of bone remodeling. Springfield (Ill.): Ch. C. Thomas 1963.

—, and D. B. HARRIS: A study of the postnatal growth of the human mandible. Amer. J. Orthodont. **50**, 25—50 (1964).

FULLMER, H. M., C. C. LINK, and M. J. BAER: A stain for bone—illustrating apposition and absorption in two colors. Stain Technol. **39**, 71—73 (1964).

GOTTLIEB, B.: Die vitale Färbung der kalkhaltigen Gewebe. Anat. Anz. **46**, 179—194 (1914).

HOYTE, D. A. N.: Alizarin as an indicator of bone growth. J. Anat. (Lond.) **94**, 432—442 (1960).

— The postnatal growth of the ear capsule in the rabbit. Amer. J. Anat. **108**, 1—16 (1961).

Massler, M., and I. Schour: The growth pattern of the cranial vault in the albino rat as measured by vital staining with alizarine red "S". Anat. Rec. 110, 83—101 (1951).

Mednick, L. W., and S. L. Washburn: The role of the sutures in the growth of the brain-case of the infant pig. Amer. J. phys. Anthrop. 14, 175—192 (1956).

Moore, A. W.: Head growth of the macaque monkey as revealed by vital staining, embedding, and undecalcified sectioning. Amer. J. Orthodont. 35, 654—671 (1949).

Myers, H. M.: Absorption studies with hydroxyapatite and alizarin. J. oral Ther. Pharmacol. 1, 165—174 (1964).

Payton, C. G.: The growth in length of the long bones in the madder-fed pig. J. Anat. (Lond.) 66, 414—425 (1932).

— The growth of the epiphyses of the long bones in the madder-fed pig. J. Anat. (Lond.) 67, 371—381 (1933).

Schour, I.: Measurements of bone growth by alizarine injections. Proc. Soc. exp. Biol. (N.Y.) 34, 140—141 (1936).

—, and M. M. Hoffman: Effects of alizarine red S injections on the teeth and bones of macacus rhesus monkey. Proc. Soc. exp. Biol. (N.Y.) 37, 710—711 (1938).

— — B. G. Sarnat, and M. B. Engel: Vital staining of growing teeth with alizarine red S. J. dent. Res. 20, 411—418 (1941).

Weinmann, J. P., and H. Sicher: Bone and bones. St. Louis: C. V. Mosby Co. 1947.

An Evaluation of the Use of Bone Histology in Forensic Medicine and Anthropology *

D. H. Enlow, Ph.D.

Introduction

The question has often been raised as to whether or not the microscopic structure of bone can be utilized for identifications of unknown bone samples. Physical anthropologists, as well as medico-legal specialists, may encounter complete or fragmentary specimens of bone in which standard anthropometric methods provide inconclusive or questionable information. If the histological structure of such material could reveal, with reasonable accuracy, details of species, age, sex, race, and similar data, the worker would have available a useful and valuable laboratory tool. Bone is a tissue of particular interest, because the hard tissue of the skeleton is normally well preserved and details of microscopic structure can be seen in specimens many centuries old.

The purpose of this discussion is to appraise and evaluate our status of knowledge dealing with bone histology in terms of its application to forensic medicine and to physical anthropology. The basis for interpretations from such observations will be described, and the particular circumstances in which conclusions are justified will be discussed. It is also of basic importance that the investigator be aware of those specific instances where an analysis of bone tissue structure *cannot* provide conclusive information relative to age, sex, species, and the many other complex variables associated with bone as a tissue. The basis for these variables will be described, and the variety of circumstances that have a direct bearing on the validity of such identifications will be defined and evaluated. The complexities of bone tissue and its growth processes are such that misinterpretations of a critical nature can be made if the observer is not well informed on the multitude of basic details dealing with osteogenesis, remodeling, and comparative histology. The objectives of this report are to evaluate the various laboratory uses of bone histology and to discuss the limitations involved. A comprehensive catalogue providing descriptive data of structural characteristics relative to species, age, and regional variations in different parts of a bone, etc. would be beyond the scope of this paper. The basic principles and concepts pertinent to the subject, however, will be reviewed and discussed.

Textbook Descriptions of Bone Tissues

Histology as a separate field had its beginnings in the early part of the Nineteenth century, and the first textbooks devoted exclusively to this subject appeared

* This research was supported (in part) by U.S.P.H.S. Grant No. DE-01903.

during the middle of this same century. Many of these texts described the typical structure of bone as a three-layered tissue composed of inner and outer circumferential lamellae enclosing a middle zone of Haversian systems (Fig. 1). This particular description has since been passed on in virtually all modern textbooks. It is often implied that all bones, from all species and at all ages following birth, conform to this particular pattern of organization. It is not generally realized, however, that this traditional plan of structure is only infrequently found, and that several other basic structural patterns are more commonly observed in the various bone tissues of different species and at different ages. Because other patterns of structure are not included in standard textbook descriptions, they remain largely unknown to all but a few specialists.

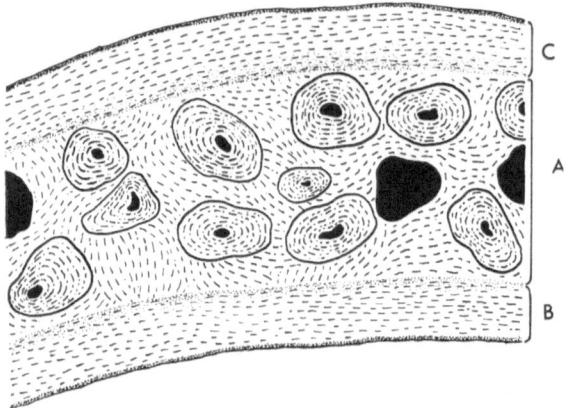

Fig. 1. This is the traditional pattern of bone tissue organization involving a three-layered cortex. In the particular section seen here, Zone A was produced by a process of cancellous compaction. This occurred when the area previously occupied a location in the metaphysis near the epiphyseal plate. Endosteal growth, by cancellous compaction, brought about an inward shift of the entire cortex present at that time. Superimposed secondary osteones are also present. Zone B was laid down as the area became progressively relocated away from the metaphysis. Continued inward growth resulted in a layer of inner circumferential lamellae, since cancellous trabeculae were not present in this area. Following reversal, periosteal Zone C was deposited as the cortex (now located in the diaphysis) increased in diameter in proportion with the bone's overall growth in length

Although a great deal of structural variation exists in the microscopic organization of bone, such differences are constant in their occurrence and result from a known series of growth and remodeling processes. With an understanding of these growth processes, the histologist can interpret the developmental basis for most of the differences in structural patterns encountered during examinations of random sections of bone tissue. It is important to realize that the histological appearance of bone differs markedly in (a) different bones of the same skeleton; (b) different parts of the same bone; (c) different areas of the same section; (d) different species; (e) and at different ages. Such differences are a direct result of specific remodeling processes. It is possible to reconstruct the developmental history of each bone, in detail, since past growth stages are recorded in the hard matrix of the bone itself. The worker familiar with the nature of the various structural differences occurring in bone can utilize them, within the limitations described later, in the identification and interpretation of unknown bone tissue samples.

Basic Types of Bone Tissue

Interpretations of bone structure and the remodeling processes that lead to it are dependent on the recognition of the different kinds of bone tissue. Since a description of the variety of basic tissue patterns is not included in standard histology textbooks, a brief review of the types of bone tissue likely to be encountered in routine studies and in random specimens of bone is presented below.

Haversian Bone

The secondary osteone (Haversian system) is the traditional structural "unit" of compact bone (Fig. 1). Contrary to widespread assumption, however, the osteone is totally absent in the bone of a great many vertebrate species. Such non-Haversian types of bone tissue include any one or a combination of the other varieties out-

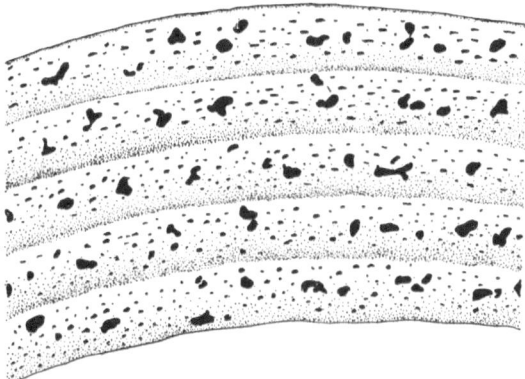

Fig. 2. *Laminar bone.* A series of broad, circumferentially arranged layers (laminae) are present. Each contains a row of primary vascular canals

lined below. In some species, including the human, secondary osteones do occur in certain specific parts of the cortex, and the distribution, placement, and density of such Haversian systems are subject to several growth and physiological circumstances. The secondary osteone develops by a process of cortical remodeling involving the resorptive enlargement of any pre-existing vascular canal. This is followed by the deposition of a cylinder of concentric lamellae within this previously formed resorption space. When Haversian systems occur in the bone of any given species, it is noteworthy that they develop in certain specific and predictable locations. Their formation, further, is directly associated with age. These factors, to be discussed in more detail in later paragraphs, can be utilized in the evaluation of unknown samples of bone.

Primary Vascular Bone

The most prevalent type of bone in most vertebrate species and at most ages is *non-Haversian* in nature. Such bone contains a series of *primary* canals that anastomose throughout the cortex (Figs. 2, 3, 5). These canals are not surrounded by the familiar cylinders of Haversian lamellae, and they have not undergone secondary remodeling changes, as described in the previous paragraph. Primary canals are oriented in a predominantly longitudinal direction, and they are found to a greater or lesser extent in almost all vertebrates at some time during postnatal skeletal development. They represent the chief type of vascular channel in human bone during the actively growing period following birth. Only in certain specific locations do primary canals become replaced by Haversian systems during the period of active skeletal growth. It is an enigma that this most basic of all bone tissue types is not mentioned or described in standard textbook descriptions of bone.

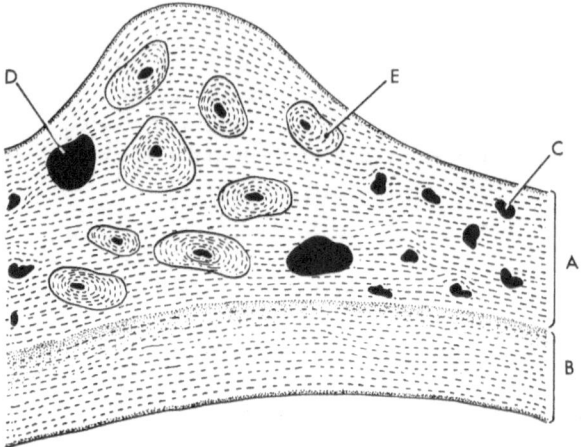

Fig. 3. In regions of muscle or tendon attachment, secondary osteones form by a process of internal Haversian remodeling. Pre-existing *primary canals* (C) become enlarged into *resorption canals* (D), and subsequent deposition of concentric lamellae within such resorption spaces results in the formation of typical *Haversian systems* (E). An inner zone of *non-vascular* lamellar bone (B) is also present in this section. Similar avascular layers can also be produced by subperiosteal bone growth

Plexiform Bone

Most primary vascular canals are oriented parallel to the principal axis of the bone. Plexiform bone, however, is a kind of primary bone tissue in which an extensive plexus of primary canals is organized in several planes throughout the cortex (Enlow, 1963). Longitudinal, radial, and circumferential canals form a symmetrical, three-dimensional, regularly arranged network (Fig. 4). Plexiform bone is characteristically found in certain vertebrate groups (particularly the artiodactyls, including the cow, sheep, deer, etc.), and it is therefore useful to the comparative histologist in the identification or interpretation of bone samples. This bone type is formed during periods of relatively rapid growth involving rather large amounts of new bone deposition.

Non-Vascular Bone

Most varieties of both primary and secondary bone involve a system of vascular channels that anastomose throughout the substance of the compact bone. Blood vessels become directly incorporated into the cortex as new deposits are laid down upon the various inner and outer surfaces of the bone. In most vertebrate forms, however, some areas of compact bone are usually present that are totally non-vascular in structure (Fig. 3, zone B). Such bone tissue is often formed during periods of relatively slow growth. Non-vascular areas of bone are more commonly encountered in the somewhat older individual, or in any region where *local* growth has become slowed. Such avascular bone usually occurs as a distinct zone in combination with other zones containing greater or lesser numbers of vascular canals. The actual number or density of canals is apparently proportional to the rate of bone deposition. Certain vertebrate groups (lizards, snakes) have skeletal tissues all of which characteristically lack vascularization in the periosteal zones of the cortex.

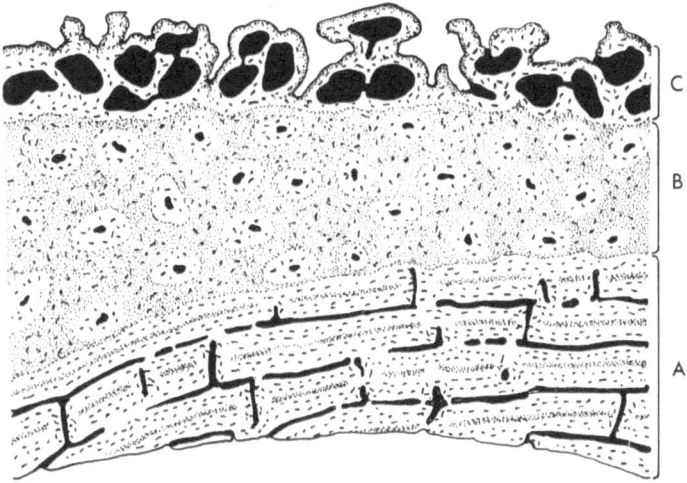

Fig. 4. Inner Zone A, which is of periosteal origin but which has become translocated to the inside of the cortex due to increased diametrical growth, is composed of typical *plexiform bone*. The primary canals in this bone type are arranged in a regularly oriented, three-dimensional plexus. The mode of formation of this type of bone parallels that of Zone B, described below. Zone B is a periosteal layer containing cylindrical *primary osteones*. The interstitial bone located between these longitudinally oriented structures is non-lamellar (woven-fibered) bone tissue. As *non-lamellar, fine-cancellous* bone is formed by sub-periosteal deposition, as in Zone C, the relatively small cancellous spaces (compare with the coarse-cancellous bone in Fig. 5) are subsequently filled with either lamellar or parallel-fibered bone. Primary osteones result. The particular types of bone tissue seen in Zones A, B, and C are characteristically produced during rapid skeletal growth.

Non-Cellular Bone

Most of the various kinds of compact and cancellous bone tissues contain variable numbers of component osteocytes. However, the bone of a great many teleost fishes is characteristically acellular in structure. Such bone can be readily recognized in routine tissue examinations.

Lamellar Bone

During bone formation, the intercellular matrix in many species is ordinarily deposited in the form of thin, successive sheets or lamellae (Figs. 3, 5). Lamellation is a result of differing orientation of the collagenous fibrils in each of the super-imposed lamellae.

Non-Lamellar Bone

Although lamellar bone represents the mature type of tissue in the human skeleton as well as in most other mammals, non-lamellar bone tissue is also an important and widely distributed variety in certain circumstances. It is formed extensively during age levels that involve a particularly rapid deposition of bone or in those specific parts of a bone where local growth is proceeding rapidly (Fig. 4). Lamellar and non-lamellar bone types are often seen in layered combinations, since growth and remodeling processes usually result in continuously changing differences in localized rates of growth (Fig. 5).

The fibrous matrix of non-lamellar bone may be either "woven" or "parallel-fibered" in composition. Woven-fibered bone is characterized by an intercellular matrix having a coarse, reticulate appearance. The woven-appearing collagenous fibers are not laid down in a stratified series of evenly spaced lamellae. Lacunae included in this bone lack the layered, regularly positioned arrangement characteristic of lamellar bone. Instead, the more densely arranged osteocytes appear randomly scattered. This particular feature is useful in the recognition and identification of woven-fibered bone.

Parallel-fibered bone, in contrast, has a uniform, regular arrangement of lacunae, and its matrix contains fibers all of which are oriented in the same general plane. It is difficult to distinguish this bone type from true lamellar bone, and positive identification can only be made using special staining methods (SMITH, 1961) or by the use of polarized light.

Coarse-Cancellous Bone

Spongy bone, when present, occupies the medullary core, and it is formed by endosteal deposition (Fig. 5). It is not ordinarily formed by direct subperiosteal apposition. The trabeculae of coarse-cancellous bone may have either a lamellar or a non-lamellar matrix.

Fine-Cancellous Bone

This type of bone tissue, intermediate in porosity between compact and coarse-cancellous bone, is widely distributed in the prenatal skeleton (Fig. 4). It occurs also in the growing postnatal skeleton in any local part of a bone where regional growth is particularly rapid. Fine-cancellous bone is typically composed of non-lamellar tissue, either woven or parallel-fibered in nature. The formation of this type of bone provides relatively large quantities of new bone tissue in short periods of time. Unlike coarse-cancellous bone, it can be of periosteal as well as endosteal origin.

Compacted Cancellous Bone

During the remodeling of a bone, the cortex in many areas grows and moves in an actual endosteal direction. This movement involves a process of direct conversion of *medullary* coarse-cancellous bone into *cortical* compact bone. Intertrabecular spaces are partially filled with new bone deposits, thereby reducing the cancellous marrow spaces to the diameter of ordinary vascular canals. The resulting compact bone has a characteristically convoluted, irregular pattern of structure, since the original trabeculae served as templates upon which subsequent deposits were laid (Figs. 1, 5).

Such compacted cancellous bone tissue is commonly seen, regardless of age or species, in most routine bone slides. If serial sections through an entire bone from proximal to distal ends are examined, more than half of the bone tissue encountered is likely to be of this widespread variety. Since standard textbooks do not call attention to it, the occurrence and the nature of this basic and important kind of bone tissue remains unknown to most workers.

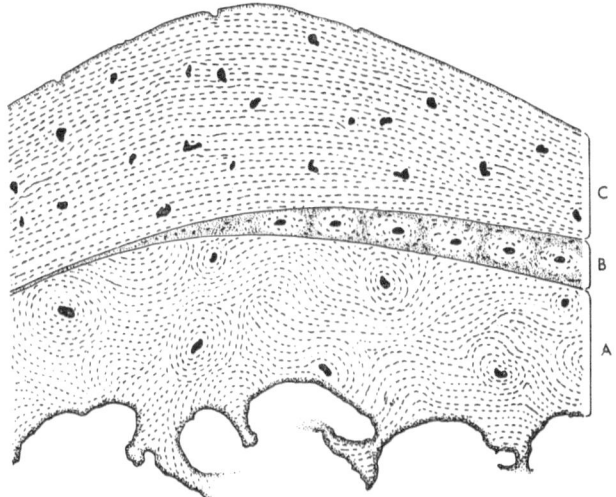

Fig. 5. Inner Zone A was formed during the inward shift of the cortex when this region previously occupied a level closer to the end of the growing bone. Zone A is composed of endosteal bone and was produced by a process of *cancellous compaction*. Note the coarse-cancellous trabeculae in the medullary cavity. As a result of its mode of formation, the lamellae and lacunae in Zone A are characteristically irregular and convoluted in their orientation. Zone B is a narrow zone composed of primary osteones and non-lamellar interstitial bone. The latter was initially fine-cancellous in structure prior to subsequent compaction. Such zones composed of single rows of primary osteones are commonly observed in the cortex of the young, growing skeleton. Zone B is of periosteal origin and was deposited after the reversal in direction of growth following the earlier formation of endosteal Zone A. It was laid down during an interval of particularly fast growth in this regional part of the bone. Periosteal Zone C is composed of *primary (non-Haversian) canals*. Note that typical Haversian systems are lacking. Concentric, encircling lamellae, as found in both the primary and secondary osteone, are not present. This is the most prevalent type of bone tissue found in the young, growing human skeleton, as well as in the bone of most other vertebrate forms

Fine-cancellous, non-lamellar bone may also be compacted. In the fetal skeleton this bone type ordinarily retains its cancellous structure, even in cortical areas of the bone. Fine-cancellous tissue in the postnatal skeleton, however, undergoes compaction soon after its formation by additional deposits of new bone within each of the cancelli. Cylinders of either lamellar or parallel-fibered bone are laid down in the tubular cancellous spaces between the woven-fibered trabeculae, thereby producing a distinctive kind of structure termed a "primary osteone" (Fig. 4). The primary osteone differs from the more familiar secondary osteone (Haversian system) in that (1) it does not result from secondary remodeling processes involving the resorptive enlargement of original primary vascular canals, and (2) it is not surrounded by an encircling resorption (cement) line. The interstitial bone located between these primary osteones is characteristically composed of woven-fibered, non-lamellar tissue representing the original fine-cancellous trabeculae prior to the compaction of its relatively small spaces. Cortical bone composed of such primary osteones is typical in age levels involving fast skeletal growth, since both woven-fibered as well as fine-cancellous bone types are involved. Such bone is found in very young human bone as well as in the young skeleton of other mammalian species. Zones or layers containing such a tissue type are often seen in combination with other kinds of tissue as a result of the complex sequence of growth and re-modeling changes that occur during the growth of postnatal skeleton. Plexiform bone, mentioned in a previous paragraph, is structurally similar to this variety of bone tissue. The two types of bone differ primarily in the orientation of com-ponent vascular canals. In plexiform bone, the canals form a uniform, three-dimen-sional plexus, in contrast to the longitudinal course of most primary osteones.

Inner and Outer Circumferential Bone

When ordinary lamellar bone is deposited on either inner (endosteal) or outer (periosteal) surfaces, it follows the already existing contour of the older surface. Bone deposits that are formed in areas of coarse-cancellous trabeculae on the en-dosteal side of the cortex conform to the pattern of trabeculation already present. If, on the other hand, cancellous bone is not present, or if trabeculae are widely spaced as in the mid-diaphysis of some long bones, new deposits of endosteal lamellar bone are laid down in uninterrupted sheets of inner circumferential lamellae (Fig. 1). Similarly, subperiosteal layers of outer circumferential lamellae result if the periosteal surface itself has an existing surface that is relatively smooth. Inner and outer circumferential layers may be composed of either vascular or non-vascular varieties of bone tissue. Contrary to one popular concept of Haversian system formation (HAM, 1953), true secondary osteones do not form *de novo* within zones of periosteal lamellar bone. Rather, remodeling processes involving resorption and secondary redeposition are always involved in the production of secondary Haversian systems. It is important that the worker be able to distinguish the primary from the secondary osteone.

Laminar Bone

In vertebrate forms that undergo cyclic periods of hibernation or which ex-perience marked seasonal changes in feeding habits (such as the turtle or alligator), compact bone tissues show distinct seasonal banding (Fig. 2). Each band is a broad lamina of bone that is either lamellar or non-lamellar in composition, and it can be either vascular or non-vascular in nature depending upon rate of formation, as outlined in preceeding paragraphs.

Special Varieties of Bone Tissue

Some vertebrates, particularly among the fishes, have unique types of bone that are not found in other vertebrate forms. The subholostean fishes, for example, have a special and distinctive kind of bone containing characteristic "lepidosteoid" tubules (ENLOW and BROWN, 1956). The presence of these structures in any unknown sample of bone is diagnostic for this particular group. The various snappers have a special "fibriform" type of bone containing massive bundles of parallel fibers that extend throughout most of the compact bone. As previously mentioned, most of the higher teleost fishes characteristically lack osteocytes in their bone. Although virtually all vertebrates have local areas that are non-vascular in structure, only members of the Order Squamata (lizards and snakes) have periosteal bone tissue that is entirely non-vascular.

Remodeling Processes in Bone

Because the calcified intercellular matrix of bone is hard and incapable of interstitial growth, a special process of structural remodeling necessarily accompanies skeletal growth. This process involves sequential adjustments in virtually all areas of the growing bone in order to continuously maintain the constant shape and relationships of the bone as a whole. These remodeling changes, involving localized combinations of deposition and resorption on the various internal and external surfaces of the bone, are responsible for the marked differences in microscopic structure between different bones, different parts of the same bone, different ages, and different species. The variety of basic bone tissue types, as previously outlined, are produced as local responses to these complex processes of growth and remodeling. Before meaningful interpretations of bone tissues can be made in terms of structural and age differences, the worker must have a thorough understanding of such processes. They are described in detail in previous studies (ENLOW, 1963).

Hazards Involved in Species Identification

Several comparative studies of the bone tissues in a large number of species representing many vertebrate groups have been made (FOOTE, 1916; AMPRINO and GODINA, 1947; ENLOW and BROWN, 1956, 1957, 1958). These works all show that major groups of vertebrates demonstrate characteristic patterns of bone structure. Attempts have been made to find reliable structural features that would permit conclusive identifications of bone at the generic or species level. Such characteristics, however, have not been demonstrated. Human bone, for example, can not be recognized as such with any reasonable degree of certainty, since some other mammalian forms have a combination and an organization of bone tissue types that more or less parallel that of human bone.

On the other hand, it is justifiable to catagorize bone samples from a negative point of view, and such information can be of value in some circumstances. For example, it would be possible to determine, in many instances, that a given specimen of bone is *not* human. Bone tissues from a great many non-primate forms have a character that would serve to distinguish them, with reasonable certainty, from the bone of man. Conversely, if a given bone pattern in an unknown sample is comparable to that of human bone tissue, the observer could only say that it might be of human origin, but that it could also belong to any one of many other forms.

Even though a particular species is not ordinarily identifiable, it may be possible to place a bone specimen in some more general taxonomic group, e.g. order or class. A number of specific examples were cited in previous sections of the present

report. The bone of some forms, such as a cow, deer, dog, turtle, any bird, etc.,
can usually be distinguished from human bone with little difficulty. However,
it is important to realize that an isolated bone fragment from any unknown specimen
or from a limited, perhaps non-representative sample of bone tissue, may not prove
adequate to show enough structural features to justify valid conclusions. If the
sample is limited to cancellous bone, or if it happens to be taken from a location
near the end of a bone where the cortex is very thin, or if it is a part of any bone
characteristically possessing a thin compacta (such as the body of a vertebra),
identifications are not practicable since such bone tissues are similar in structure
among most mammals regardless of species. Bone of a reptile or a bird, in general,
might be distinguished from mammalian bone under such circumstances (ENLOW,
1957, 1967), but to attempt further more detailed identifications would not be
warranted. *Only if adequate sampling* is possible can any interpretation or identification
be justified. Even then, it may only be possible to say that a given specimen could
be of human origin, but that the bone might also belong to a bear, cat, monkey,
etc., all of which have bone tissues that may compare favorably with human bone,
if variables such as age, location, and part of bone are all considered. These latter
factors are discussed in later paragraphs.

Group Characteristics of Bone Tissue Structure

Every vertebrate species from fish to mammal, fossil or modern, has an individual
combination of bone tissue characteristics (see under "Types of Bone Tissue"
above). These structural features may be unique to any individual group, or more
frequently, they are similar in nature to one or more quite unrelated groups. The
individual combinations of bone tissue types are a direct result of the particular
circumstances of skeletal growth in any given group. Rate of body growth, body
size, special skeletal adaptations, and seasonal or environmental factors all influence
the nature of skeletal deposits (ENLOW, 1963, 1967).

Since the present paper is restricted to a brief appraisal, a systematic description
of specific bone tissue characteristics in all vertebrate groups is not presented.
Previous reports dealing with surveys of various representative vertebrate forms
(cited above) will serve to provide the investigator with a start in the evaluation
of any problem concerned with the practical identification or interpretation of an
unknown bone tissue specimen. It must be realized, however, that these studies
are necessarily rather broad in scope and that such surveys cannot provide exhaustive
information regarding any given species. In fact, no single vertebrate species,
including man, has to date received a comprehensive descriptive study of all the
individual bones, all parts or areas in each bone, and at all age levels. Although
much pertinent information is presently known about some selected forms, a
systematic catalogue of the kind just mentioned has yet to be compiled. Such a
work, particularly if concerned with human bone, would be quite worthwhile,
although it would represent a somewhat tedious project. The conceptual background needed for this kind of a study is now reasonably well understood and
described, and such a project is encouraged. Some specific human bones, particularly
those of the face (ENLOW and HARRIS, 1964; ENLOW and BANG, 1965) have already
been worked out in detail as to microscopic structural characteristics in the different

parts of each bone and at different ages. Specific, detailed information dealing with the other major bones of the human skeleton is now needed.

While existing knowledge is sufficient in many cases to distinguish major vertebrate groups if adequate samples are available, it is apparent that the current lack of any comprehensive handbook dealing with key species makes it necessary for the worker to employ "control" specimens when attempting unknown identifications. Because of the various differences in bone structure, as previously emphasized, direct comparisons with samples of known species, age, etc., must necessarily be employed. Special care must be taken in the selection of the particular bone and the specific location from which bone sections are removed for sectioning and study. These factors are outlined in the following paragraphs.

Identification of Specific Bones and Parts of Bones

Because each bone has an individual shape, the sequence of remodeling changes that accompanies growth results in different patterns of microscopic components from bone to bone. Also, because any single bone may differ in histological structure between taxonomic groups, corresponding species differences occur even when considering the same skeletal element. The arrangement of structural components in a mid-diaphyseal section of a human femur will differ markedly from a similar location in the humerus, even in the same individual. These differences in pattern all result from the remodeling changes that occur in bones of different gross shape and different overall size. Other differences in the actual kind of bone tissue present also occur, and these are based on age, rate of growth, and other such factors (see below). These latter variations, if present, are all superimposed over structural differences produced by the growth of bones each having a different morphology.

Specific location within any given bone is a consideration of fundamental importance. During the growth of any bone, each regional area of the bone is successively repositioned in its location as the entire bone increases in size. This process involves extensive remodeling in virtually all individual parts of the bone as a whole as local adjustments in shape and dimension take place. This means that no two areas of any bone will be quite alike in microscopic structure. In fact, major differences are ordinarily found in the appearance of bone sections that are prepared from different locations. Also, because lateral shifts of the cortex are usually involved in the maintenance of shaft curvature during growth, different sides of a bone will differ noticeably in structure even though they are both from the same level of the same bone. The basis for these various differences has been described in principle, and studies involving an application of these principles to specific human bones has also been made (ENLOW and HARRIS, 1964; ENLOW and BANG, 1965). It is evident that any investigator dealing with unknown bone tissue specimens must be familiar with background information dealing with bone tissue differences before meaningful interpretations of bone tissue structure on a comparative basis can be made.

As already pointed out, it may be necessary to compare unknown bone samples with known controls when attempting identifications. It is essential that the same skeletal element be used, and that the precise same location within the bone be selected. To do otherwise could result in misleading or inaccurate conclusions.

If only an isolated fragment of bone is available for study and the particular bone
or part of bone cannot be recognized with certainty, it may still be possible to
determine whether or not the specimen is of possible human origin or to place
it in some particular group. This could be done on the basis of the actual type
of bone tissue present rather than upon the pattern or combination of patterns
characteristic for single bones.

If the species of any bone sample is known, then it may be possible to identify
which bone is involved. It is also possible to determine the exact location within
the bone. In each case, a series of control preparations will necessarily have to be
used for direct comparison.

Age differences

It was mentioned in earlier paragraphs that the specific types of bone are each
adapted to different kinds of growth circumstances. Rate of growth directly in-
fluences the nature and the structure of bone deposits. Growth rate must be considered from two points of view: (1) the overall rate of body or skeletal growth,
and (2) the local rate of growth in each of the many parts of any bone. It has been
shown that fine-cancellous, non-lamellar bone is a tissue type that is associated
with particularly rapid skeletal growth, whether of periosteal or endosteal origin.
It represents the predominant tissue type in the prenatal skeleton, even in areas
of the cortex. Compact bone, as such, is absent or sparsely distributed in fetal
bones. Fine-cancellous, non-lamellar bone is also a common tissue type deposited
during the first two or three postnatal years. Following birth, however, many of
the spaces in developing fine-cancellous tissue soon become filled with either lamellar
or parallel-fibered bone. The structural result is a composite bone tissue type composed of *primary osteones*. The interstitial bone between the osteone cylinders is
usually composed of woven-fibered, non-lamellar bone representing original fine-
cancellous trabeculae. Bone tissue of the human fetal skeleton, as well as postnatal
bone up to approximately the second or third year, can thus be recognized on the
basis of the presence and wide distribution of these characteristic tissue types.

As age increases, and as the bone grows in overall size, some parts of the bone
grow a great deal faster than others. Because of such differential rates of growth,
a variety of different combinations of bone tissues accumulate in different parts
of the same bone. In any local area involving a particularly rapid increase in size,
fine-cancellous, non-lamellar bone is deposited. Primary osteones subsequently form
within the cancelli. In adjacent areas where growth is less rapid, or in the same
area at some succeeding developmental period when growth becomes slowed, lesser
amounts of bone of a different type are laid down. Circumferential lamellar bone
containing variable numbers of primary canals results. These primary vascular
canals are not surrounded by cylinders of Haversian lamellae. The number of such
canals in any given area of lamellar bone seems to be an index of the relative rate
of its formation. As the skeleton increases in size beyond the early postnatal years,
a fairly fast rate of growth continues and results in periosteal deposits having a
rich distribution of primary non-Haversian canals. As skeletal growth slows with
increasing age, the number of such canals included in new bone deposits usually
decreases. Similarly, in specific parts of any growing bone where local growth is

relatively fast, more canals are found in comparison to the less vascular bone in areas that involve slower growth. Each level of the bone, however, will continuously change its rate of growth as it becomes successively relocated into different relative positions as the entire bone increases in size. As a result, a variety of mixed types of bone are commonly seen superimposed upon each other. It is apparent that a series of different levels from one end of a bone to the other must be examined in order to properly determine the overall nature of growth in that particular bone.

When evaluating the distribution of the various kinds of primary bone in terms of age levels and growth rates, only *periosteal* deposits should be considered. The several basic types of endosteal bone are not a good index of differential growth, since inward growth is concerned primarily with remodeling adjustments in shape and diameter as the various parts of the bone become successively relocated during growth. Overall increases in diameter are largely a function of periosteal bone deposition. Since cortical zones composed of endosteal bone can occupy two-thirds or more of the total bone mass, particularly in the proximal and distal thirds of a long bone, the importance of recognizing the various kinds of endosteal bone is evident.

The presence and the nature of distribution of Haversian systems (secondary osteones) is a useful guide to relative skeletal ages. Such observations serve to supplement information gained by the evaluation of any patterns of primary bone also present. Several key considerations must be understood, however, before such determinations become meaningful.

Secondary osteones are rarely encountered in young, growing postnatal bone tissues except in certain specific areas. In very young bone, as mentioned above, fine-cancellous tissue with or without primary osteones represents the predominant bone type. Subsequent to birth and on until skeletal maturity is reached, true Haversian systems are characteristically found in only two localized regions of the cortex. First, they are found selectively in zones of the compacta that are *endosteal* in origin. In such locations, they are typically quite numerous and can occupy much of the endosteal portion of the cortex. They have been superimposed over any of the various kinds of original primary endosteal bone. A layer of inner circumferential lamellae may also be present as an innermost zone. Again, the observer is cautioned that distinction between endosteal layers of the cortex and periosteal zones is of basic importance.

The second characteristic location in the growing postnatal skeleton in which Haversian systems are routinely encountered is any specific region of compact bone, either endosteal or periosteal, that involves direct *muscle attachment*. Elevated crests and tubercles, particularly, show dense concentrations of such Haversian systems. In regions of periosteal bone that do not involve muscle anchorage, secondary osteones are absent or only thinly scattered in distribution. Former areas of muscle attachment, however, that have since become relocated into new positions which no longer involve such insertion can also show numerous secondary osteones. It has been suggested (ENLOW, 1962) that the formation of secondary osteones in either periosteal or endosteal zones of cortical bone may represent a direct response to muscle relocation on the growing bone. In endosteal deposits of bone, the inward growth of the cortex during remodeling involves an extensive resorption of bone from the periosteal surface. Also, during bone growth, tubercles, crests, and all other points of muscle attachment are constantly moving in their location in order to maintain a constant relative position on the bone as a whole. It has been suggested that the process of successive relocation and reattachment of the moving muscle insertion is accomplished by Haversian reconstruction within the cortical bone itself, thus providing new bone that is directly continuous with the connective tissue framework of the repositioned muscle.

The distribution of Haversian systems in the two general locations just described cannot be utilized by the observer in accurately assessing skeletal age, since they are present from childhood through old age. However, osteones that form in other periosteal zones after skeletal maturity is reached can be used in approximating general age levels. Only those particular osteones that form in periosteal deposits should be considered.

Prior to skeletal maturity (early twenties in man), most *periosteal* zones of the cortex, except at points of muscle attachment, have a widespread distribution of non-Haversian bone. Any one or combination of several of the basic kinds of primary bone can be present, as described in previous paragraphs. The presence of such primary bone deposits thereby provides the worker with an index as to approximate growth level. Subsequent to the period of active skeletal growth, most areas of the cortex, including such periosteal zones, become progressively replaced by secondary osteones during processes of internal cortical remodeling. This involves the replacement, over a period of time, of former deposits of primary as well as secondary bone by new generations of Haversian systems. These replacement osteones accumulate and become superimposed in complex patterns as the bone continues to age. This process of Haversian reconstruction continues on into extreme old age until most of the original bone present at the time of early skeletal maturity, including both endosteal and periosteal zones, has become replaced. The number and the extent of distribution of secondary osteones, thus, progressively increases from early maturity until death. The classic study by AMPRINO and BARAITI (1936) illustrates this progression very nicely. Some restricted areas of primary (non-Haversian) periosteal or endosteal bone can be encountered in old bone, but such deposits may represent remodeling alterations in response to changes in skeletal morphology associated with old age, including changes in weight, posture, loss of teeth, or other similar factors that can result in slight adjustments of a bone's shape and dimensions.

Because Haversian systems within endosteal zones of the cortex in the young, growing skeleton are numerous and densely arranged, as they also are in areas of muscle attachment in periosteal zones, caution must be used to assure that such Haversian bone is not mistaken for the dense Haversian tissue that is characteristic of most periosteal areas in the older skeleton. This consideration is of basic importance. The *endosteal* Haversian bone of the young skeleton and *periosteal* Haversian bone of the aged skeleton can be distinguished only by the experienced bone histologist. Since periosteal as well as endosteal zones become progressively replaced by Haversian bone with advancing age, it becomes increasingly more difficult to accurately differentiate between such zones.

The width of the cortex may also become noticeably thinned at advanced age levels, which provides another characteristic useful in age estimations. Further, the various canals and cancellous spaces of aged bone can become enlarged as a result of senile osteoporosis.

The foregoing account refers specifically to human bone. Certain other species or taxonomic groups follow a similar age progression, in principle, but because body size, growth rate, and a number of other metabolic factors may differ, corresponding differences in the time sequence and the extent of Haversian distribution may also occur.

Sex Differences and Individual Variations

Significant differences between sexes in the microscopic structure of bones are not evident among known mammalian forms. Quantitative differences in cortical thickness and dimensions of bony ridges or tuberosities may be present, but such values are based on statistically large numbers of individuals. Any attempt to assign sex to an unknown bone sample, considering only its histological structure, would not be justified in view of racial and age differences in the various dimensions of any given bone. Among egg-laying vertebrates, quantitative changes are known to occur in medullary areas of a bone, but this would be of questionable value in any practical sense when attempting to analyze unknown bone tissues.

Variations do occur in the histological patterns of bone tissue structure between individuals. Such differences, however, are restricted largely to quantitative variations, since the basic patterns of structure themselves have been found to be fairly constant between individuals as well as between right and left sides of the same individual. Marked differences can be encountered, it will be recalled, if sections from any two bones are not examined from *precisely comparable* levels. Because quantitative differences can occur between bones of different individuals, the exact placement of the various microscopic zones will vary accordingly. Conclusions, therefore, must be arrived at on the basis of a series of sections prepared through any given area in order to account for slight differences in the actual dimensions of microscopic patterns and the extent of transitions between them. The observer is concerned primarily with the nature of the patterns themselves and the various combinations of patterns, and he must be certain that a range of microscopic sections are examined in any part of a bone in order to have an accurate overall picture. Slight differences in the exact placement of the patterns relative to each other are of no particular concern. However, since such variations can mislead the worker into assuming the patterns themselves are different, he must employ serial sections to account for such variations.

Actual variation in the character of any bone tissue pattern itself, however, may also be present. In studies of the bones in the human face (ENLOW, 1965) it has been found that a few specific but limited areas have a range of predictable variations in their histological patterns of structure. It has not yet been determined whether such variations are based on racial differences, but it is apparent that they can have an effect on gross morphology in the areas concerned.

Environmental Influences on Bone Tissue

The bone of homothermic vertebrates has not been found to have any particular relationship with known seasonal, climatic, or geographic factors. Nutritional deficiencies, of course, can result in a variety of pathologic conditions directly affecting bone, and these must be considered if involved in any bone specimen being studied. Some poikilotherms, however, have bone tissues that are directly affected by seasonal changes or by cyclic differences in feeding habits. In these forms, such as the turtle and alligator, the cortex is distinctly laminated. Each lamina represents a kind of growth ring produced during successive periods of body growth. Comparative age estimations are possible between individuals, but actual age determinations are not possible. Because the nature of cortical banding reflects environmental conditions that existed at the time of formation, a study of laminar bone can be utilized in appraising such factors (ENLOW, 1967).

Use of Bone Histology in Anthropology

Advances in our understanding of the basic biological principles governing organic evolution have given much insight into the nature and meaning of evolutionary mechanisms associated with the phylogeny of human bone as a tissue. Although a thorough, systematic account of bone tissue phylogeny in general, including a survey of all major vertebrate groups, is beyond the scope of the present report, several basic concepts bear directly upon the application of bone histology to anthropological studies. These are briefly outlined in the following paragraph.

Evolution is customarily considered in terms of organs and organ systems. Phylogenic studies of *tissues* have not, in general, been studied in depth. Our present understanding of the nature of tissue evolution is based largely on studies that have been made from a "horizontal" point of view in the sense that only contemporary forms are considered. In such studies, individual forms are traditionally ranked as "higher" or "lower" in a phylogenetic scale. The principle of orthogenesis is usually emphasized in order to link these species into a single-line evolutionary progression. A series of species are selected and then arranged into a vertical lineage assumed to represent an actual phylogenic scale from primitive to advanced. Each such form that is selected for this scale is a living, modern species, and all are chosen in order to provide a more or less constant, gradual series of stages that conform to an ascending range of tissue advances, considering the particular structures being studied.

This approach is hazardous and has very little real meaning, since direct lineage between such contemporaneous forms did not occur. The arrangement of the different species or groups necessarily represents an artificial sequence of stages, and the inference that the living species making up this scale represent modern descendents of original ancestral conditions is based upon presumption. To rely upon such an approach can lead to erroneous conclusions and will serve to obscure other perhaps more meaningful interpretations.

Some evolutionary studies of tissues have utilized the Biogenic Law in attempts to relate phylogenic significance to fetal and adult tissue transitions. For example, in some texts and references, it is stated that the ontogeny of mammalian bone recapitulates the phylogeny of bone tissues in the lower, non-mammalian forms. It is held that adult human bone, composed of densely arranged Haversian systems, represents the phylogenic culmination of all bone tissue evolution. The non-Haversian nature of the non-lamellar bone in the human fetus is regarded as a direct recapitulation of comparable bone tissue types in some adult non-mammalian forms (such as the frog). However, a key point of basic significance which is not considered is that the complex variety of different kinds of bone tissue each represent an adaptation to specific functional factors. Variations in the microscopic structure of bone between different vertebrates is largely a result of the individual circumstances of growth and the sequence of remodeling changes that take place in the growing bones of each species or group. The various combinations of the basic types of bone tissue in any species, in any particular bone or part of a bone, and at any age level are all determined by the rate of overall body growth, rates of localized growth in different areas of the same bone, the particular shape of the bone, and the overall size of the skeleton. These factors all contribute to the presence and nature of the distribution of lamellar, non-lamellar, fine-cancellous, plexiform, etc., types of bone tissue. This assortment of bone tissue types is found in varied combinations in virtually all bony vertebrates, and they can be considered as phylogenic only in the sense that they have become genetically established within the framework of overall body growth in each individual group. The presence of bone types in many adult mammals that are similar to the bone in adult forms of lower vertebrate groups is, in itself, sufficient cause to discredit this older concept.

Bone and the other calcified tissues are the only parts of the body that permit histological examination of fossil tissues. For this reason, only in the skeleton can a tissue be studied from a true vertical approach. Tissue preservation is usually adequate for study in most fossils, even in the more ancient geologic periods. The question can be properly asked as to whether or not the microscopic structure of bone can be utilized in detailed studies of human phylogeny.

Comparative studies of bone tissue, including both fossil and modern forms and considering a wide variety of groups (loc. cit.), indicate that bone histology can be of value in studies of evolution only if a *functional* approach is used. Conclusions or interpretations based solely on comparative similarities and differences in histological structure are not warranted. It would not be profitable, for example, to select a series of pre-human types and to then examine random bones in the hope of finding a sequence of progressive, straight-line structural changes in the organization of the bone as a tissue. The phylogeny of bone, in contrast, conforms to a principle that involves a direct response of microscopic structure to the specific circumstances of growth and remodeling which occur in any given species. It would not be possible to demonstrate a graded series of species-specific patterns each of which is uniquely characteristic for a given pre-human type in an ascending lineage of presumed transitional forms. It would not be possible to prove, using bone histology, that any particular hominoid form is directly ancestral to man, or to any species believed to be within the human "line" of evolution. It would not be possible, further, to prove that any given form is a member of any particular human or subhuman genus or species solely on the basis of bone tissue structure. Bone histology, in addition, would be of no value in attempts to judge any paleoecological conditions at the time a fossil specimen was actually living. Sex, moreover, cannot be determined. Age, however, might be estimated using the same procedure as that described for modern human bone if proper precautions for both sampling and interpretation are followed, as outlined previously.

Although these limitations can reduce the usefulness and versatility of bone structure in some specific kinds of anthropological studies, a great deal of basic, meaningful information may nevertheless be obtained. It is emphasized again that the microscopic structure of bone is a result of individual growth and remodeling processes in each species. Since different groups have markedly different characteristics in many gross skeletal features, the specific remodeling circumstances that lead to these differences can be studied and usually explained by detailed examinations of the bone tissues involved. Because past remodeling changes which occurred during the growth of any bone become recorded in the substance of the bone itself, it is possible to retrace, in detail, the sequence of these changes. Thus, the worker familiar with basic mechanisms of bone tissue remodeling can recognize microscopic patterns and tissue combinations produced by the complex variety of remodeling changes, and he can thereby accurately reconstruct the growth history of that particular bone.

This procedure is particularly useful when applied on a comparative basis in order to evaluate the actual basis for differences in skeletal morphology between any two groups or species. For example, the unique human chin undergoes a series of remodeling changes that differs noticeably from the corresponding region in anthropoids (ENLOW and HARRIS, 1964). The simian shelf, also, shows similar differences in histological structure between man and non-human primates. The entire human face, further, demonstrates a complex pattern of remodeling adjustments during growth. Although the facial bones, particularly the maxilla and mandible, appear to grow in a forward and downward direction, growth actually takes place in a predominantly posterior course. The entire bone is simultaneously relocated in a forward direction as actual bone deposition occurs along most posterior free surfaces. As such growth proceeds, however, bone resorption occurs at the same time over almost all of the anterior surfaces of the mandible, premaxilla, and the maxilla, including its zygomatic processes. The zygomatic process of the maxillary bone as well as the separate zygomatic bone itself actually grow and move in a posterior course as the entire maxillary bone grows in a corresponding direction. This movement of the zygomatic complex functions to maintain its constant position in the growing maxillary bone as a whole.

The presence of an anterior maxillary surface that is almost entirely resorptive in character appears to be associated with facial recession and the decreased prognathism characteristic of man. Resorptive remodeling involved in the alveolar bone superior to the mental protuberance of the human mandible also appears to be related with this same circumstance. It is suggested that the upright stature of man may be involved in these growth characteristics. Such a shift in posture necessitates corresponding adjustments and changes in the position and orientation of many skeletal elements, including those of the face. Thus, facial regression, as revealed by microscopic patterns of bone tissue, may be involved in the adaptation of the human face to an erect position, as well as to an increased cranial capacity. The reorientation of the orbits and the frontal bone in their relative positions, the vertically oriented face, the protruding nose, and the corresponding lack of protruding and nasal-obstructing jaws may all represent interrelated, functional adaptations associated with upright posture and a greatly enlarged cranium.

A detailed study of the comparative situation in a number of vertebrate forms, including primates as well as representative species from other groups, is now needed in order to relate variations in facial growth patterns with body stature and facial morphology. Such studies can provide a profitable approach to phylogenetic interpretations of human skeletal characteristics based on actual functional relationships between bone, gross morphology, and body growth. Such a study could be both horizontal and vertical, depending upon availability of specimen material. In addition to the various adaptations of the face, as just outlined, a number of other body areas have yet to be considered from such a point of view. For example, the human pelvis, the occipital condyles, the foramen magnum, the mastoid tuberosities, and certain long bones may all show remodeling changes related with posture and other human morphological characteristics. Such studies are encouraged.

Bone histology may also be utilized in studies of the ontogenetic changes that occur during the growth of certain skeletal areas, including the facial complex. Progressive remodeling changes in different parts of the same bone as well as between different but closely associated bones are known to produce corresponding changes in the proportions, dimensions, and relationships of these bones. For example, the characteristic changes that take place in the gross appearance of the human face with increasing age are based on the underlying remodeling movements of the facial bones.

Summary

The applications of bone histology in appropriate areas of forensic medicine and physical anthropology were appraised and discussed. The limitations and special hazards involved in such interpretations were outlined and described. The worker should realize what cannot reasonably be accomplished in forensic studies involving bone tissues and recognize those circumstances where bone histology has no direct value. This report is intended primarily as an evaluation rather than as a comprehensive presentation of descriptive data.

Textbook descriptions dealing with the postnatal structure of bone are misleading and for the most part do not provide a representative or accurate picture of the *typical* appearance of bone as a tissue. In the present report, the basic variety of commonly seen patterns of bone structure are briefly described, since accurate recognition of the different kinds of microscopic bone tissue patterns is essential for any utilization of bone histology in forensic medicine and anthropology. Significant differences in the structure of bone between different bones, different parts of the same bone, different ages, and different species are pointed out. The developmental basis for these differences is briefly outlined.

The characteristics of bone which can be utilized in identifications of unknown bone samples are summarized. Features that permit the localization of a bone specimen as to specific position within the whole bone are also briefly reviewed. Progressive changes occurring with increasing age are evaluated and structural characteristics related to age differences are discussed. In all such identifications, the investigator must be well informed as to the complexities of bone tissue structure and the marked structural changes which occur during progressive bone remodeling. A background knowledge limited only to a casual understanding of bone structure or one which does not go beyond standard textbook descriptions can result in inaccurate conclusions or misleading interpretations.

Bone histology can provide the anthropologist with a great deal of basic information provided the cautions outlined above are observed. Bone histology is useful primarily in functional studies of specific skeletal adaptations to the physical form of the species.

References

AMPRINO, R., and A. BAIRATI: Processi di recostruzione e di riassorbimento nella sostanza compatta delle ossa dell'uomo. Z. Zellforsch. **24**, 439—511 (1936).

— — Le variazioni nella struttura dell'osso in relazione all'età e la loro importanza medico-forense. Arch. Antrop. crim. **16** (1), 61—74 (1938).

—, and G. GODINA: La struttura delle ossa nei vertebrati. Comment. Pont. Acad. Sci. **11** (9), 329—467 (1947).

ENLOW, D. H.: Principles of bone remodeling. Springfield (Ill.): Ch. C. Thomas 1963.

— Functions of the Haversian system. Amer. J. Anat. **110** (3), 269—306 (1962).

— Bone. In: The biology of the reptilia [E. E. WILLIAMS, C. GANS, and A. BELLAIRS (eds.)]. London: Academic Press 1967 (in press).

—, and S. BANG: Growth and remodeling of the human maxilla. Amer. J. Orthodont. **51** (6), 446—464 (1965).

—, and S. O. BROWN: A comparative histological study of fossil and recent bone tissues. Part I. Tex. J. Sci. **7** (4), 405—443 (1956).

— — A comparative histological study of fossil and recent bone tissues. Part II. Tex. J. Sci. **9** (2), 186—214 (1957).

— — A comparative histological study of fossil and recent bone tissues. Part III. Tex. J. Sci. **10** (2), 187—230 (1958).

—, and D. HARRIS: A study of the postnatal growth of the human mandible. Amer. J. Orthodont. **50** (1), 25—50 (1964).

FOOTE, J. S.: A contribution to the comparative histology of the femur. Smithson. Contr. **35** (3) (1916).

HAM, M. B.: Histology. Philadelphia: J. B. Lippincott Co. 1953.

KERLEY, E. R.: The microscopic determination of age in human bone. Amer. J. Phys. Anthrop. **23** (2), 149—163 (1965).

SMITH, J. W.: Collagen fibre patterns in mammalian bone. J. Anat. (Lond.) **94** (3), 329—344 (1961).

WEINMANN, J. P., and H. SICHER: Bone and bones, 2nd ed. St. Louis (Mo.): C. V. Mosby Co. 1955.

Evaluation of Skeletal Impacts of Human Cadavers *

H. R. Lissner ** and **V. L. Roberts**

In the study of the effect of impact on the human body at injury-producing levels it has been necessary to use models because of the danger of severe and serious damage to living volunteers. Since accurate information regarding the response of the human body to injury-producing impacts is negligible, it is apparent that available anthropometric dummies are of no value in such investigations. Because of the very significant differences between impact responses in animals and in humans it is difficult to justify the extrapolation of responses in impact experiments with live animals to similar responses in humans. This leaves the human cadaver as the only reasonably good model available for such experimental work, results of which can only be evaluated in terms of the behavior of the skeletal system. But even with this list of available results, embalming and the presence of dead tissue instead of living tissue would be expected to have some effect on the response obtained to a specific impact.

In order to attempt to evaluate the significance of these factors a series of impact experiments were performed on living anesthetized dogs. The response was determined by means of acceleration measurements and direct strain measurement on the bones using electric strain gages. These tests duplicated a series of experiments previously made on human cadavers with impact accelerations applied to the spine in the caudocephalad direction. After completion of the tests the animals were sacrificed and embalmed. From three to seven days later the embalmed animals were subjected to the identical testing program to which they had been subjected while living.

The procedure employed and results obtained from the use of human cadavers has been previously described (HODGSON et al., 1963). The experimental procedure and results obtained with the use of living and embalmed dogs will be described in this paper. The factors indicating differences in behavior in the living and dead animals will then be applied to the measured response of the human cadaver, in order to predict the behavior of the living human body when subjected to the same impact conditions.

Dogs were used in the experiments, not because they were considered to be ideal from the standpoint of anatomy, but because it was necessary to perfect the required techniques. Now that the techniques have been perfected more expensive animals, e.g. monkeys, will be used for later experiments. From comparison of

---

* This research was supported by U.S. Public Health Grant No. AC-00054-06, Division of Accident Prevention.
** Deceased.

the response of the embalmed dog with that of the living animal the authors believe that the results obtained when using primates will compare very favorably with those resulting from the use of dogs.

A total of 30 mongrel dogs was used in the conduct of these tests. In all tests the impact was applied to the ischial tuberosities in a caudocephalad direction at a constant acceleration of 18 G for a duration of 170 milliseconds. The time rate of change of applied acceleration (jerk) was varied from 250 G to 3000 G per second to determine its effect on the skeletal strain response. The apparatus used in these tests has been described elsewhere (Roberts and Lissner, in press) and is the same as that used previously in similar tests with human cadavers.

Surgical Preparation and Application of Gages to the Test Animal

After anesthetizing the animal with nembutal, the right brachial vein was catheterized. The catheter terminated in a fitting for attaching a syringe through which additional nembutal could be administered as required. A tracheotomy was performed on the animal in order to assure proper ventilation throughout the test.

In order to apply strain gages to the anterior aspect of the fifth or sixth lumbar vertebrae, it was necessary to make an incision along the midline of the abdomen. After displacing the abdominal viscera to one side, the spine was exposed and a section of periosteum was removed from the vertebra to which the strain gage was to be cemented. The area was cleaned and degreased with acetone, following which one or more strain gages were cemented in place with cyanoacrylate (Eastman 910) cement. These gages are paper thin and are available in sizes as small as 0.397 mm in length. After application of the gages, to which lead wires of No 36 stranded copper insulated with teflon had previously been soldered, the entire installation was coated with a water-proofing material. The lead wires were sutured to tissue in the immediate area of the gage in such a fashion that tugging on the external leads could not apply tension to the gage itself. A large loop of the lead wires was left in the Abdominal cavity and the wires were brought out through the incision. The cavity was sutured in multiple layers until it was closed with the wires protruding from the closure.

Gages could not be applied to the anterior aspect of the thoracic vertebrae because of difficulties involved in going through the chest. Therefore, the vertebrae were exposed by an incision along the back, and gages were mounted on the liminae and on the spinous processes of the first and second thoracic vertebrae. An aluminum plate designed to hold the Glennite accelerometer outside of the incision was also screwed to the process of the first thoracic vertebra. The gage was applied and insulated in the same fashion as described above for the lumbar vertebrae installations. A large loop of the lead wires was left in the incision and the wire sutured again outside of the incision.

Because of the extreme flexibility of the spinal column of the dogs in the sagital plane it was necessary to provide some restraint. This was accomplished by placing the animal in a light box and filling the area between the animal and the wall of the box with a very light foaming plastic (polyurethane). The plastic was applied to the level of the thorax and the animal's head was held erect by means of a fitting holding the nose. The dog, thus, maintained its vertical position when accelerations were applied in the caudocephalad direction. The leads from the strain gages and from the accelerometer were brought out through the restraining device and connected to the recording system. Because of the severity of the surgical procedures, the animals with gages applied to the lumbar vertebrae did not have gages applied to the thoracic vertebrae and vice versa.

This procedure permitted tests on the living, anesthetized dog following which the animal could be sacrificed, embalmed and left in the foaming plastic with all gages intact until further testing was done on the embalmed animal. Thus, we were able to maintain the dog in exactly the same alignment for tests in the embalmed and living condition.

Tests performed on all 30 of the dogs were made with an acceleration of 18 G's being applied to the container and the animal. The jerk during these tests was the only variable. Total time and terminal acceleration magnitude were held constant but the jerk was varied from 250 to 3000 G per second. After a level of 18 G was reached, it was maintained for the remainder of the stroke. In order to check the reproducibility of the tests and fracture of the vertebrae, the tests were begun at low jerk levels and gradually increased until the maximum was reached. Following this, the test at low jerk level was repeated in order to check whether the original results were reproduced.

When all the tests on the living anesthetized animal had been completed, the animal was sacrificed by the administration of a large dose of nembutal. In the interim between sacrificing and embalming, the carcass, still in the box, was kept in a cold storage room. The embalming in all cases was completed within 24 hours after sacrificing the animal. Most of the animals were tested within 72 hours after embalming, but in four of the tests the animals were kept for as long as a week before undergoing the test.

The results of the tests in which the animal was kept for the longer period before testing did not vary appreciably from those obtained with the animals that were tested shortly after embalming. The dog was embalmed by supplying embalming fluid through the brachial catheter as well as by direct injection of the fluid into all areas of the body including the brain. This was necessary because of the relative inaccessibility of the major blood vessels of the animal when enclosed in the foaming plastic.

After the test series on the embalmed dog was completed, the animal was removed from the box and x-rayed in order to verify the location of the gages with respect to the vertebrae on which they were applied and to check for possible fractures which might have occurred during testing. In some cases the body cavity was opened to determine the reason for malfunction of some of the gages. This procedure also indicated that the embalming process was quite complete as was evident by lack of deterioration of the internal organs of the body.

Results

Figs. 1, 2, and 3 are reproduced from an earlier publication (ROBERTS and LISS-NER, in press) to help illustrate some of the results of the experiments. They show data taken from particular dogs indicating typical results obtained from these tests.

Because of variations in size of the animal and of the vertebrae, gage locations, etc. direct comparisons among all animals tested were difficult to make.

Figs. 1—3 indicate the kind of results obtained with all animals, and also the spread of data between the in vivo and cadaver results with any one animal. The ratio between the strain at a constant acceleration and that value obtained initially due to the particular jerk applied is defined as the *dynamic load factor*.

Fig. 1 illustrates the relationship between the dynamic load factor and jerk for both living and embalmed animals with strain gages mounted on the anterior aspect of the sixth lumbar vertebra. This is a typical example of the results obtained in this test sequence showing a dynamic load factor of less than 1.5 in both the living and embalmed animals with an extremely close relation between the cadaver and the in vivo results. In other words, the effect of death and embalming on the

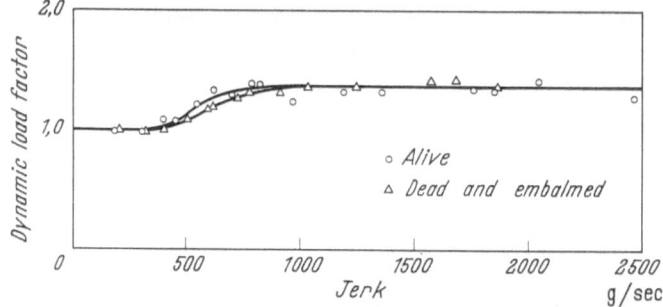

Fig. 1. Plot of dynamic load factor versus jerk for a strain gage mounted on L_6 of a dog. From Roberts and Lissner (in press)

response of the lumbar vertebrae was relatively insignificant. The maximum difference obtained with gages mounted on the lumbar spine was greater than that shown in Fig. 1, but in all the animals studied in the lumbar spine sequence, the difference between the cadaver and in vivo results did not exceed 20% in the flat portion of the curve past the transition region. It is significant that in all instances the dynamic load factor, as a function of jerk, was higher for the living dog than for the embalmed animal. This factor must be taken into account in applying the results obtained from dogs to the tests of human cadavers.

Another factor was that the flat portion of the curve was reached sooner in the living animal than in the dead one and occurred in the neighborhood of 700 to 750 G's per second. In the case of the embalmed animal the flat portion of the curve was not reached until 750 to 1250 G's per second was achieved. The differences shown in this example are typical of those obtained from the strain gages in both the lumbar and thoracic, regions and from the accelerometer in the thoracic region. Mean strain values of 1000 micro-millimeters or more were consistently obtained from all the gages.

It is worth pointing out that these results agree with the reported values of dynamic load factor in the human cadaver (Fig. 4) where the value was less than 1.5 for the lumbar vertebrae as shown by the strain gage readings. Fig. 2 shows the strain data obtained from the gages mounted on the first and second thoracic vertebrae of a dog. The data are consistent with other data obtained in that the maximum dynamic load factor was found to be a value of 1.7 or greater with the living animal. The living animal again shows a higher dynamic load factor than that obtained for the embalmed animal. The gage data obtained from the thoracic

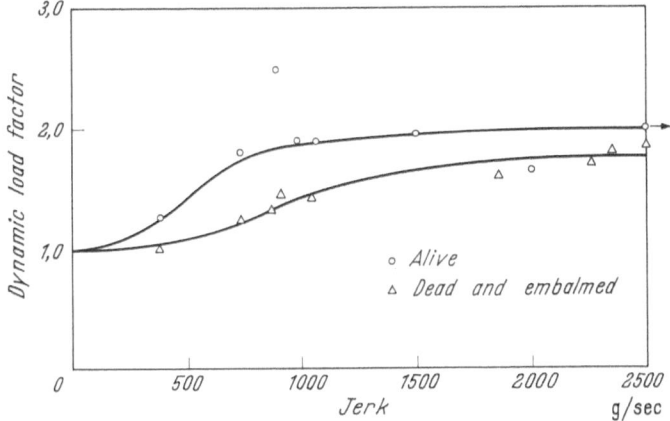

Fig. 2. Plot of dynamic load factor versus jerk for a strain gage mounted on the spinous process of T_2 of a dog. From ROBERTS and LISSNER (in press)

area indicated that the platcau in the curve of dynamic load factor versus jerk occurred at a somewhat higher level than in the lumbar region. For the living animal the plateau occurred in the range of 750 to 1250 G's per second while in the embalmed animal it had a range of 1000 to 1500 G's per second. The difference between

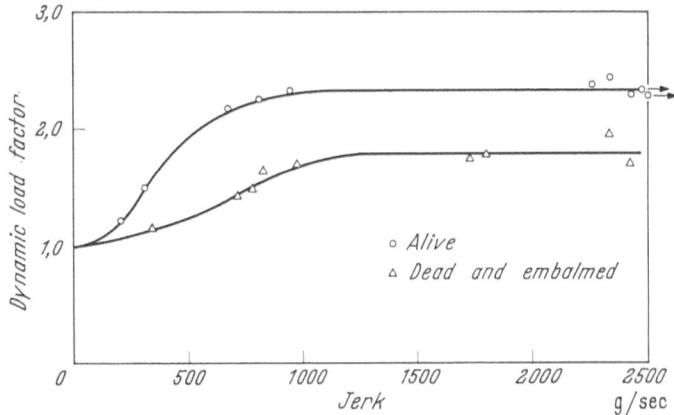

Fig. 3. Plot of dynamic load factor versus jerk for an accelerometer mounted on the spinous process of T_1 of a dog. From ROBERTS and LISSNER (in press)

the results for living and embalmed animals did not exceed 20% in the flat portion of the curve although in the rising portion of the curve the difference is more pronounced.

Fig. 3 indicates the results obtained from the Glennite accelerometer mounted on an aluminum plate screwed to the first thoracic vertebra. The general shape of the curve is the same as the curves of strain shown in Figs. 1 and 2. The accelerometer data, however, tended to be more erratic and to show more scatter than those obtained from the strain gages. This is probably because of the location of the accelerometer which was mounted on a cantilevered projection so that it was actually outside of the body. The dynamic load factor varied in magnitude from 2 to 3 as indicated by the accelerometer. The results from the embalmed animal generally were lower than those obtained with the living animal. The results seemed to remain within the 20% maximum spread in the flat portion of the curve. In the living animal the rising portion of the curve occurred earlier, with acceleration readings between 250 and 750 G's per second jerk, than in the embalmed animal, in which acceleration readings ranging from 500 to 1000 G's per second jerk were found.

In the performance of these tests it was found that the location of the strain gages, particularly on the thoracic vertebrae, is crucial with respect to the mean strain level attained during acceleration. Since the spinous process is loaded in bending, gages placed on the middle of the process, when viewed laterally, had a mean strain of almost zero. This yielded unnaturally high values of dynamic load factor which had little significance as far as injury was concerned. When gages were placed to one side or the other of the spinous process, however, consistent results were obtained with mean strain in the area of 500 micro-millimeters per millimeter.

Discussion

The results of the tests obtained from the 30 dogs show that a reasonable correlation exists between the cadaver and the living animal, and that the cadaver is a reasonably good model for the behavior of the living test subject. The weight distribution of the cadaver is the same as for the living biological system and the over-all mechanical response of the cadaver seems to approximate quite closely that of the living system with regard to the behavior of skeletal structures.

The data show, however, that care should be taken when comparing cadaver data, in the transition range or rising portion of the dynamic load factor vs. jerk curve, with those obtained from living animals because results in this range exhibit significant differences between the response of the cadaver and that of the living animal. The differences in the characteristic curves of dynamic response of the living and dead animals indicate that living tissue and bone have a lower spring rate than that of the embalmed tissue and bone. This is shown by the different shapes of the two curves in the transition zone between a dynamic load factor of one and maximum or terminal values of this load factor. Living tissue reaches the maximum amplification sooner while the stiffer system of the embalmed material requires a higher jerk to reach the terminal value. This is precisely the behavior predicted mathematically (PAYNE, 1962) for the response of elastic systems subjected to ramp inputs of the type used here.

The results previously obtained with human cadavers (HODGSON, LISSNER, and PATRICK, 1963) are shown in Figs. 4, 5, and 6. The actual data are represented by the solid lines while the corrected data are shown by the dashed lines. Although the data have been presented previously, they are shown here to aid in understanding the results of the present study. Figs. 4 and 5 show the strain-dynamic load factor response of lumbar and thoracic vertebrae while Fig. 6 shows the acceleration-dynamic load factor response of the sternum, all with respect to rate of onset.

These results are representative of those obtained with all the cadavers tested and were chosen because of the relative similarity, with respect to location, between the vertebrae studied in the dog and in man. The dynamic response of the dog

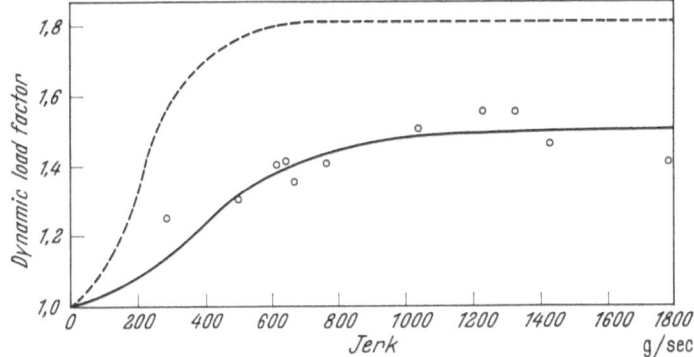

Fig. 4. Plot of dynamic load factor versus jerk for a strain gage mounted on L_4 of a human cadaver. The solid line shows the recorded data while the dashed line indicates the predicted results for the living human

and man show some very similar features. The recorded values of dynamic load factor at similar locations agree quite closely in magnitude if the 20% spread in

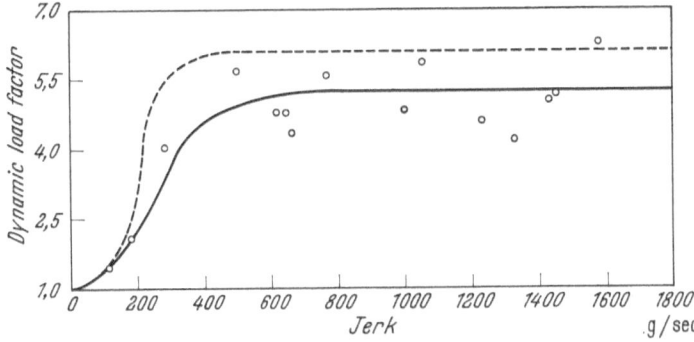

Fig. 5. Plot of dynamic load factor versus jerk for a strain gage mounted on the spinous process of T_5 of a human cadaver. The solid line shows the recorded data while the dashed line indicates the Predicted results for the living human

the data is taken into account. The point beyond which no further amplification occurs due to rate of onset is different for dogs and humans. This, however, is not surprising when the difference in mass of the two subjects is considered.

The increases in dynamic load factor shown in projected results are based upon the maximum of 20% spread observed at all levels in the dog tests. The more rapid rise in the load factor-jerk curve projected for living human tests is based on consistent findings of this nature with animals. This reflects the lower spring rate observed with living subjects.

The data show that with higher values of dynamic load factors a low mean strain of 500 micro-millimeters per millimeter could still result in peak strain levels high enough to cause fracture. Cadaver results should be regarded as a lower limit and not be applied directly to living subjects as tolerance limits.

The projected results for living human subjects are susceptible to the greatest possible errors in the transition zone between no amplification and maximum amplification. It is in this region where the cadaver does not provide a good dynamic model. The over-all pattern of dynamic response is good, however, and the shape of the dynamic load factor-jerk curves are basically the same for the living and the dead and embalmed systems.

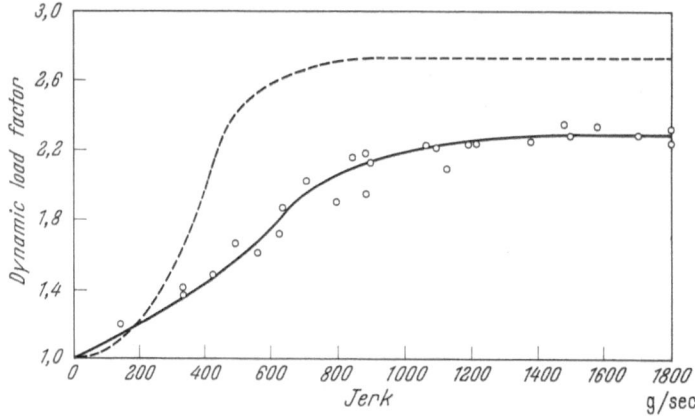

Fig. 6. Plot of dynamic load factor versus jerk for an accelerometer mounted on the sternum of a human cadaver. The solid line shows the recorded data while the dashed line indicates the predicted results for the living human

Conclusions

From the data obtained in this study the following conclusions can be drawn:

1. The cadaver may be used as a model for the strain response of living biological systems with a high degree of confidence.

2. The strain response of living systems is greater than that of embalmed systems and will show decreased sensitivity to increased rate of onset (jerk) at lower levels.

References

Hodgson, V. R., H. R. Lissner, and L. M. Patrick: Response of the seated human cadaver to acceleration and jerk with and without seat cushions. Hum. Factors 5, 505—523 (1963).

Payne, P. R.: The dynamics of human restraint systems. Proc. of the Symposium on Impact Acceleration Stress, Publ. 977, National Acad. Sci. Washington, D. C. 1962.

Roberts, V. L., and H. R. Lissner: A correlation between cadaver and in vivo results. Eight Stapp Car Crash and Field Demonstration Conference. Wayne State University Press (in press).

The Tensile Properties of Single Osteons Studied Using a Microwave Extensimeter*

A. Ascenzi, E. Bonucci and A. Checcucci

In the last twenty years increased attention has been devoted to problems concerned with the mechanical properties of bones. The works in this field deal with macroscopic bone samples, i.e., whole skeletal segments or machined specimens but direct investigations on microscopic specimens are lacking.

Recently an extensive research program on the mechanical properties of single osteons was started in our laboratory and in a previous paper the results of an investigation on the ultimate tensile strength of the same units were given (Ascenzi and Bonucci, 1964). The aim of this paper is to present the preliminary and tentative results of a quantitative investigation on the tensile deformation of single osteons from human and ox femoral shafts using a purposely devised apparatus.

Plan of the Investigation
Introduction

As reported by Koch (1917), elasticity is the property a material has of returning to its original dimensions and shape when the external forces producing distortion are removed.

Every solid body is elastic within certain limits and, according to Hooke's Law, the deformation produced in it is directly proportional to the force producing it and to the internal force resisting it. If the force producing the deformation exceeds a certain limit, the distortions will increase more rapidly than the forces, and when the load is removed the body will not entirely regain its former shape and dimensions, but will have a permanent set. The maximum unit stress up to which the deformation increases proportionally to the stress, is called the elastic limit. This limit varies with the kind of material and the kind of stress — tension, compression and shear.

Young's modulus of elasticity is the ratio *unit stress/unit deformation* and varies with the material and the kind of stress. Generally the modulus of elasticity used is that for tensile stress and is indicated by the letter E.

If P is the total tensile stress in kilograms, S the area of cross section in square centimetres, l the original length of sample in centimetres and d the total deformation in centimetres, then

$$E = \frac{\text{unit stress}}{\text{unit deformation}} = \frac{\dfrac{P}{S}}{\dfrac{d}{l}} = \frac{P\,l}{S\,d}.$$

* This work was supported by a grant of the National Research Council of Italy.

In these units E is expressed in kilograms, per square centimetre. The modulus of elasticity is constant, with very few exceptions, for any given material for stress below its elastic limit but after passing the elastic limit it steadily decreases. Working unit stresses, or stresses which the body may carry constantly or repeatedly without damage, should always be well below this limit.

Literature

The first investigation on tensile properties of bone was carried out by Wertheim (1847) who used samples prepared from the shaft of human femurs and fibulae. The author observed that only in the case of dry bone was there a relation of direct proportionality between tensile elongation and load applied to the specimen. In fresh bone Hooke's Law was not strictly respected, the elongation being somewhat greater than one would expect.

According to Carothers et al. (1949), the works of Rauber (1876) and Messerer (1880) are among the most ambitious efforts made in the 19th century to evaluate the mechanical properties of bone by actual tests. However, according to Smith and Walmsley (1959), Rauber investigated only the tensile deformation of bone and stated that Young's modulus averaged 2.90×10^6 lb/sq.in. for dry femur and 2.67×10^6 lb/sq.in. for dry tibia.

More recently Hallermann (1935) attempted to analyse the tensile elongation of samples machined from ox compacta. He emphasized that bone elongation increased proportionally to the applied tensile load until the ultimate tensile strength was reached. Therefore, Hallermann concluded that bone was an entirely elastic material. In 1945, Marique tested under direct tension a specimen machined from a presumably dry human femoral shaft. A somewhat curve-line relationship between elongation and tensile loading was found. The modulus of elasticity ranged from 183,962 to 200,000 kg/cm².

Dempster and Liddicoat (1952) tested samples of compact human bone machined from a number of femurs, tibiae, and humeri under tension in an engineering testing machine. Stress-strain curves for dry and for wet compact bone were given by a first straight part indicating a proportionality of stress and strain in line with Hooke's Law. Beyond the upper proportional limit each later increment in stress caused excessive yielding and permanent plastic deformation. In both the straight part of the curve and in the deviating plastic region greater strains were produced in wet bone than in dry by a given stress. In this connection dry bone had a modulus of elasticity in tension ranging between $2.69 \pm 0.420 \times 10^6$ and $2.86 \pm 0.612 \times 10^6$ lb/sq.in. whilst the modulus of elasticity of the wet bone dropped to an average value ranging between $1.73 \pm 0.128 \times 10^6$ and $1.77 \pm 0.232 \times 10^6$ lb/sq.in.

The tensile strength of human bone was carefully analysed by Evans in a series of papers published in the 1950's (see Evans and Lebow, 1951, 1952; Evans, 1955, 1958) as well as in his very interesting book entitled "Stress and Strain in Bones" (1957). In the first group of these investigations the tensile properties of 242 specimens of compact bone from the femurs of adult male cadavers were determined. All specimens were of a standardized size and were tested by approved engineering techniques. Half of the specimens were air-dried at room temperature and tested dry, whilst the other half were placed in a physiological saline solution and tested wet. On drying the average tensile strength of the specimens increased but their percentage elongation under tension fell. The samples tested wet, which probably more closely approximate the properties of living bone, had a greater percentage elongation under tension. This is also evident from the shape of the stress-strain curve which is a straight line to fracture for dry samples but a curve for wet specimens. The specimens from the middle third of the femoral shaft had the greatest average tensile strength and energy-absorbing capacity. The age of the individual seemed to have little influence on tensile strength.

Similar comparative studies on all the long bones of the inferior limb revealed that the tibia had the greatest average tensile strength and the fibula the greatest percentage elongation under tension. The femur was the weakest in these respects. As in the femur the middle third of the tibia and fibula had the greatest average tensile strength. The middle third of the tibia also had the greatest percentage elongation but in the fibula, and also in the femurs, the proximal third of the bone had the greatest elongation.

The microscopic structure from tested bone specimens furnished evidence that, in general, the larger the number of small osteons and minute fragments the lower the tensile strength. Moreover the tensile strength of bone is greater when the collagen fibres of the osteons and their remnants are predominantly parallel to the direction of loading.

In 1959 SMITH and WALMSLEY investigated the elasticity of bone in bending and in tension and suggested a procedure for the determination of the value of Young's modulus for this tissue. Evidence was furnished that the deformation of bone under stress varies with the duration of stress, with the fluid content of the specimen, and with its temperature. Only Young's modulus in bending was determined.

Material and Method

As discussed in a previous paper (ASCENZI and BONUCCI, 1964), it appears very difficult at the present time to isolate samples corresponding to a whole osteon in order to test tensile properties. Such a bone unit is well recognizable under the microscope only when the tissue is reduced to a transparent section of a maximum of 50—60 micra, i.e., a thickness much less than the average diameter of osteons in human and ox bone (180—200 micra). For this reason the present tensile strength studies were made on samples corresponding to portions of longitudinally sectioned osteons.

Longitudinal sections of femoral shafts, ranging from about 20 to 50 micra in thickness, were prepared by grinding on glass plates. Particular attention was paid to avoid heating the material. The samples were obtained from osteons sectioned through their longitudinal axis applying the dissection technique described by ASCENZI and FABRY (1959) and ASCENZI and BONUCCI (1964).

In this way it was easy to prepare samples with the shape seen in Fig. 1. Here only the middle portion of half a longitudinally sectioned osteon appears totally isolated and shows parallel boundaries. Both ends of the sample enter into square lugs in order to assure a good fixation in the tensile apparatus. The totally middle portion of the sample had a length corresponding to the distance between the jaws of the tensile apparatus, i.e., 0.4—0.6 mm.

The cross-sectional area of the sample at the middle was used in calculating ultimate tensile strength and modulus of elasticity. In this connection thickness was evaluated accurately by mounting the sample edgewise under the microscope and using an eyepiece micrometer.

The samples were tested in a wet and dry state, respectively, to determine the influence of moisture on the tensile behaviour. The cross-sectional areas were measured in both states.

The tested osteons revealed some structural peculiarities in the degree of calcification as well as in the structure of the matrix as regards the orientation of the collagen bundles. The degree of calcification was determined by microradiographic technique. Among the different arrangements resulting from the fibre bundle direc-

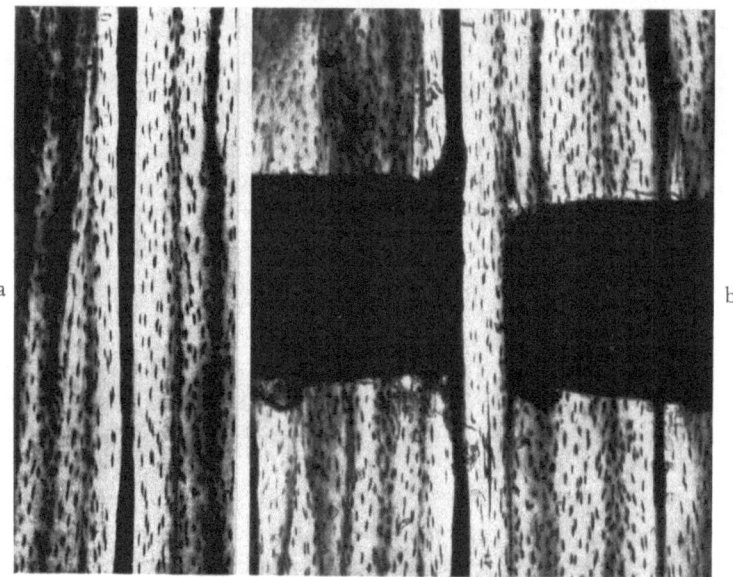

Fig. 1a and b. a Longitudinal section of femur compacta from a man of 30 showing an osteon along the mid-line. b A specimen ready to be tested has been obtained from the right half-portion of the same osteon. Polarizing microscope. ×50

tion in successive lamellae, those characterizing two types of osteons were chosen. In the first one the fibres have a marked longitudinal spiral course with the pitch of the spiral changing so slightly that the fibres in one lamella make an angle of nearly 0 degrees with the fibres of the next one. Under the polarizing microscope the osteons belonging to this type appear homogeneously bright in longitudinal section. In the second type the fibres in one lamella make an angle of nearly 90⁰ with the fibres of the next one. Under the polarizing microscope the osteons of this second type reveal an alternation of dark and bright lamellae in longitudinal section.

In some instances the samples were decalcified by treatment in a versene solution buffered to pH 7. The decalcification was checked by measuring the increase in birefringence at regular intervals of time (cf. Ascenzi and Bonucci, 1961). The minimum time required for decalcification averaged about 2 hours and 30 minutes.

The samples were subjected to a direct tensile force parallel with their long axis. For this purpose the ends of each specimen were fixed in especially designed jaws. One jaw (the upper one) was fixed. A thin nylon thread was attached to the other jaw and charged with loads of 1 g each. The elongations of the osteon were recorded with a purposely devised microwave extensimeter (see later).

Particular attention was paid to ensure that the tensile force was applied along the long axis of the samples, avoiding any torsion of the sample itself. All the measurements were performed at a temperature of about 20^0 C.

The material used in the present research was obtained from femoral shafts of oxen and two human subjects 30 and 80 years of age. As far as is known, no pathological bones were included in the tests. The total number of prepared osteons was 91, of which 17 were discarded as unsuitable for measurement. The remaining 74 units were made up of 21 ox osteons and 53 human osteons. The dry osteons were obtained by keeping the bone at room temperature, the wet ones by hydration of the material using saline solution or distilled water.

To conclude, the successive steps of the technique may be summarized as follows:

(1) Preparation, by grinding, of longitudinal sections, 20—50 micra thick from femoral shafts.

(2) Recording of microradiographs from the same sections.

(3) Selection of osteons for measurement of tensile strength according to the degree of calcification as well as the direction and distribution of the collagen bundles as revealed by the polarizing microscope.

(4) Isolation of the osteons.

(5) Measurement of tensile properties.

The Microwave Extensimeter

The apparatus used for measuring the variations in length of osteons subjected to tensile stress is a microwave extensimeter based on cavity and pulse techniques. Its informing principle may be summarized as follows.

Let h and D, respectively, be the height and the diameter of a cylindrical metallic cavity functioning as a resonator for electromagnetic waves. The resonant frequency f_0 of this cavity will depend on h, D and the configuration of the electromagnetic field inside the cavity (mode). If the type of configuration chosen is that commonly indicated by "Te_{012} mode"* the following relation holds (cf. Montgomery, 1947)

$$f_0^2 = \frac{C^2}{D^2}\left(1.488 + \frac{C^2}{h^2}\right) \tag{1}$$

C being the velocity of light in vacuum.

Any variation in the height h of the cavity produces, of course, a variation in the resonant frequency. By measuring this variation it is possible to deduce, by using (1), the corresponding variation in h.

Fig. 2a is the diagram of the cavity used for measuring the elongation of the osteon specimens submitted to tensile stress. The upper end of a specimen is fixed in a jaw. The lower end is fixed in another jaw supporting a disc of plexiglas, whose

* As will appear evident later the Te_{01n} mode was chosen because: (a) the functioning of the cavity in this mode does not require electrical contact between the lateral and the end walls of the cavity, so that the height of the cylinder can be varied; (b) the presence of a dielectric thread running along the axis of the cylinder (see later) does not disturb the functioning of the cavity, because for Te_{01n} modes the electric field is null along the axis.

lower metallized surface forms the upper plane of the cylindrical cavity. The disc may roll into the vertically orientated cavity and bears a very thin nylon thread to the center. The thread runs along the axis of the cavity and leaves by a small hole at the center of the lower fixed plane. When traction is applied to the thread by attaching some weights to it the osteon sample undergoes an elongation which makes the upper plane of the cylindrical cavity fall, thus reducing the height of the cavity. Because the elongation of the sample (Δl) and the reduction in the

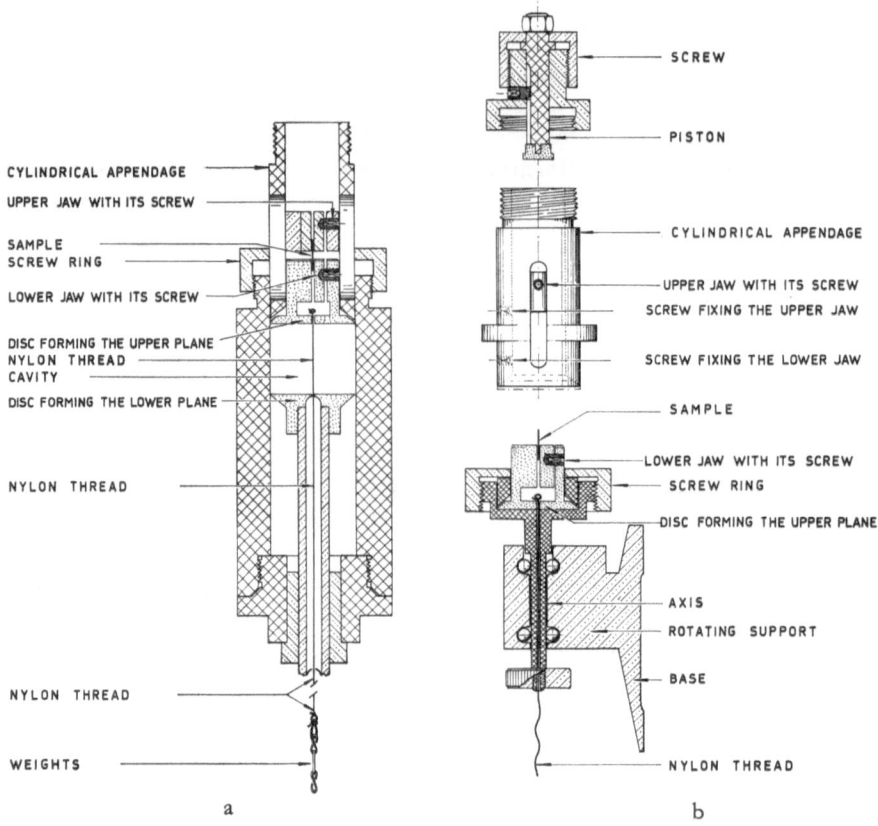

Fig. 2a and b. a Diagram of the cavity used as extensimeter. b Diagram illustrating the technical details concerning the fixing of the osteon sample to the apparatus

height of the cavity ($-\Delta l$) are of exactly the same magnitude the variation in length of the sample is easily deduced from the corresponding variation in the resonant frequency of the cavity. If Δl is very small compared with the height h of the cavity, equation (1) gives the corresponding fractional change in the resonant frequency

$$\frac{\Delta f_0}{f_0} = -\frac{\Delta l}{h} \left(\frac{1}{1 + 1.488 \, (h/D)^2} \right). \tag{2}$$

Under the present experimental conditions, i.e., $D = 2.22$ cm; $h = 1.78$ cm; $f_0 \approx 23,400$ Mhz, one may deduce that a variation in the sample length of 10^{-7} cm = 10 Å produces a fractional change in the resonant frequency of an order of magnitude of 5×10^{-8}. As the ultimate sensitivity of the apparatus is $\Delta f/f \approx 10^{-8}$, one can detect changes in the length of the sample as small as a few Angströms. In the case of osteon samples the elongation produced by an applied force amounting to 1 g varies from about 1,000 to 50,000 Å and the ultimate tensile load ranges between 20 and 120 g according to the nature and dimensions of the specimens. This means that the measurements require a sensitivity lower than that furnished by the apparatus. Therefore a high degree of accuracy is guaranteed.

As has been pointed out the problem is that of measuring small fractional changes in the resonant frequency of a cavity. For this purpose a pulse technique was used which presents the advantage of being relatively simple and assuring high sensitivity and accuracy by the elimination of all spurious effects produced by the fluctuations of the frequency and power of the oscillator (see later). This technique was developed at the Physical Institute of the University of Pisa, for spectroscopic purposes, by GOZZINI and co-workers. Here a merely qualitative description concerning the functioning of the apparatus will be given. More detailed information can be found in the paper of BATTAGLIA et al. (1958).

The resonator cavity is coupled to two waveguides. Through the first one the cavity is fed by a Klystron oscillator with a frequency very close to 23,400 Mhz. The other waveguide ends with a matched crystal detector connected to a device for the interpretation of the transmitted signal, i.e., an oscilloscope. The frequency of the Klystron is varied by superposing an isosceles triangle tension on the continuous tension of its reflector. This means that its frequency increases linearly from a minimum value f_1 to a maximum value f_2 and afterwards decreases linearly from f_2 to f_1. The modulation follows a periodic law at a frequency of 50 cps. In other words, the frequency of the electromagnetic wave at the input of the cavity varies linearly from f_1 to f_2 during $1/100$ sec and from f_2 to f_1 during the next $1/100$ sec. The cycle is renewed 50 times each second. If f_1 and f_2 are chosen in such a way that the frequency of the cavity f_0 is intermediate between them, the cavity itself will act as a filter of the electromagnetic radiation generated by the Klystron, i.e., the crystal detector will be fed with radiation having a frequency equal or quite close to the resonant frequency of the cavity. The power transmitted from a cavity depends on the frequency. If the power P_0 furnished by the oscillator is constant in the range from f_1 to f_2, the power of the crystal detector will be

$$P_t = k P_0 \frac{1}{1 + 4 Q^2 \left(\frac{f - f_0}{f_0} \right)^2} \tag{3}$$

k being a constant of proportionality depending on the diameter of the holes by which the cavity is coupled to the waveguides, f the frequency of the wave and Q the quality factor of the cavity.

As the oscillator frequency is a linear function of time and the voltage V_c at the crystal terminals is proportional to the incident power, V_c will vary with time according to the same law (3), i.e.,

$$V_c = V_0 \frac{1}{1 + 4 \frac{Q^2}{f_0^2} v^2 (t - t_0)^2} \tag{3'}$$

where t_0 is the instant when the Klystron frequency attains the value f_0 corresponding to the resonant frequency of the cavity, and v is the velocity of the frequency modulation. $v = \pm \frac{f_2 - f_1}{t_2 - t_1}$ is positive during the half period, when the Klystron frequency increases with time, and negative during the next half period, when the Klystron frequency decreases.

From (3') it is seen that during each half period the voltage at the crystal terminals reaches a maximum V_0 at the instant in which the Klystron frequency corresponds to f_0.

The tension is applied to the input of a linear amplifier whose output feeds a pulse discriminating circuit capable of supplying 1 μ sec voltage pulse when V_c reaches the highest value V_0. In this way the pulse characterizes the instant when the Klystron frequency reaches the value f_0, corresponding to the resonant frequency of the cavity. Obviously pulses are produced twice a period, i.e., during the half-period when frequency is increasing and during the next half-period when frequency is decreasing.

Besides the device described here, the same Klystron feeds a second channel identical to the first one, i.e., composed of a transmission cavity (resonating at the frequency f_0'), a crystal detector and a discriminator. During each period of the frequency modulation the latter circuit gives two pulses revealing the instant at which the Klystron frequency reaches, at each half-period, the value f_0', corresponding to the resonant frequency of the cavity. It follows that, at each modulation half-period of frequency, the delays between the pulses coming from the two discriminating circuits are:

$$\Delta t^+ = \frac{f_0 - f_0'}{v}, \quad \Delta t^- = \frac{f_0 - f_0'}{v}. \tag{4}$$

where the positive sign characterizes the delay between the pulses produced during the half-period in which the Klystron frequency increases and the negative sign characterizes the delay at the next half-period in which the frequency decreases. As the modulation velocity v is of opposite sign during the two half periods equation (4) shows that the two delays are of opposite sign. This means that the order of succession of the pulses at a half-period is reversed at the next one.

Let us use the pulses supplied by the first discriminating circuit to trigger the sweep of a synchroscope and let us send the second circuit pulses, conveniently delayed, to the synchroscope vertical deviation plates. Then one will observe, on the synchroscope screen, two pulses whose "apparent" temporal distance is

$$\Delta t = \Delta t^+ - \Delta t^- = \frac{2(f_0 - f_0')}{v}.$$

A frequency variation Δf in the cavity induces the following distance variation

$$\delta(\Delta t) = \frac{2\,\Delta f}{v}. \tag{5}$$

In the present apparatus the interval $f_2 - f_1$ is 10 Mhz and is covered in $^1/_{100}$ sec. It follows that $|v| = 10^9$ hz/sec. Introducing this value in (5) one may deduce that a 1 μ sec variation of the "apparent" temporal distance of the pulses on the synchroscope screen is produced by a 500 cycles variation of the resonant frequency of the cavity, i.e., by a fractional change $\Delta f/f \approx 2 \times 10^{-8}$. This last corresponds to a variation in length of 4 Å in the osteon sample (2).

The sensitivity of the apparatus may be changed by suitably varying the modulation velocity v. The sensitivity is reduced by the fact that the position of pulses on the synchroscope screen has a certain fluctuation. This jetter is due to the noise voltage at the amplifier output, mostly produced by the crystal detectors. The fluctuation causes indeterminateness in the frequencies discriminated by the pulse generators.

The sensitivity is characterized by the ratio $(\Delta f/f)$ min. According to apparatus theory, this ratio can be expressed as

$$\left(\frac{\Delta f}{f}\right)_{min} = \frac{4\sqrt{3}}{9}\frac{\eta}{Q} \approx \frac{\eta}{Q}$$

η being the ratio between the noise voltage and the signal voltage at the amplifier output. As in the apparatus $\eta = 5 \times 10^{-5}$ and $Q \approx 10^4$, one may deduce $(\Delta f/f)$ min $\approx 5 \times 10^{-9}$.

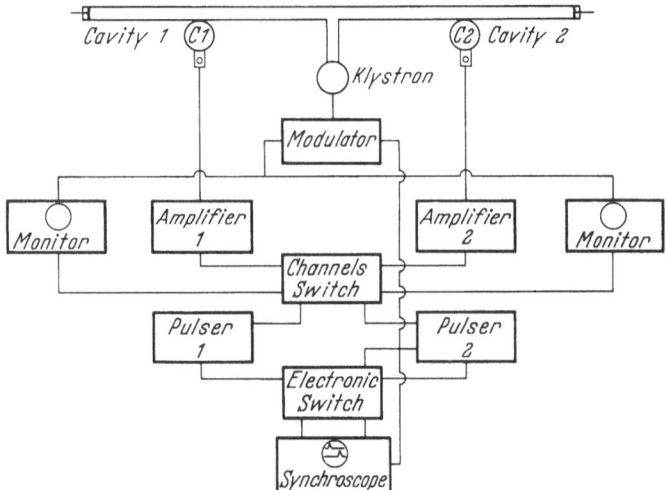

Fig. 3. Block diagram of the extensimeter

In practice mechanical vibrations and other causes reduce the sensitivity to an order of magnitude of about 5×10^{-8}. In this way it is possible to record variations in length of osteons reaching ten Å. This is a very high sensitivity when one considers that the single recorded elongations ranged between 1,000 and 50,000Å.

The block diagram of the apparatus is given in Fig. 3. Fig. 4 shows the whole apparatus as operated by us.

The variations in length of the samples were calculated from the corresponding variations in the "apparent" temporal distance of the pulses on the synchroscope screen, using the already mentioned formulae and substituting in them the values corresponding to the resonant frequency of the cavity f_0, the modulation velocity v, the diameter D, and the height h of the cavity.

A direct control of the apparatus was also carried out using two different methods. In the first the bone sample was replaced by a thin calibrated tungsten wire having a well-known modulus of elasticity to tension and showing, of course, an exactly predictable elongation when a well specified charge was applied to it.

The second method consisted of introducing into the control cavity dried air at known pressure. This procedure induces a fractional change in resonant frequency, according to the formula $\Delta f/f = (n-1)$, where n is the refractive index of the air (being $n-1 = 270 \times 10^{-6} \frac{p}{760}$ and p indicating the air pressure in mm Hg).

These two control methods have furnished similar results with a tolerance of 5 per cent. The apparatus having worked at a relatively low degree of sensitivity, the pulses on the synchroscope screen remained still and their position was determinable with an error lower than 5 per cent.

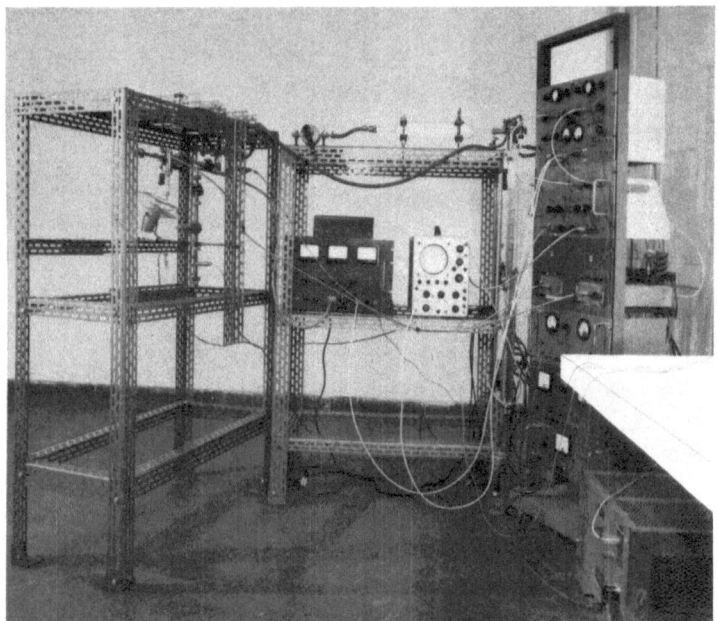

Fig. 4. Photograph of the whole apparatus

Therefore the absolute measurements of length were accurate to within 5 per cent and measurements concerning variations in length for a single sample were accurate to within 1 per cent.

A greater accuracy is not necessary for absolute measurements of length, because it would be impossible to determine with an accuracy of the same order of magnitude the geometrical dimensions (length and section) of the osteonic sample (see Ascenzi and Bonucci, 1961, 1964).

Technical Details for Fixing Samples in the Apparatus

To fix a sample in the jaws, the disc which forms the upper plane of the cavity must first be taken away, then, using a screw ring, fixed firmly in a support able to rotate about the central axis (Fig. 2b). One end of the sample is fixed in the lower jaw, where its alignment can easily be controlled under the microscope by axially rotating the system.

When the sample is in place the screw ring is removed. The cylindrical appendage containing the upper jaw is lowered round both disc and lower jaw, and held firmly by replacing the screw ring. The lower jaw is fixed to the appendage by tightening its fixing screw.

At this point a device incorporating a piston is screwed onto the upper end of the appendage in order to control the forward movement of the upper jaw to the desired level round the sample. When this has been done the upper jaw is fixed to the appendage by tightening its fixing screw.

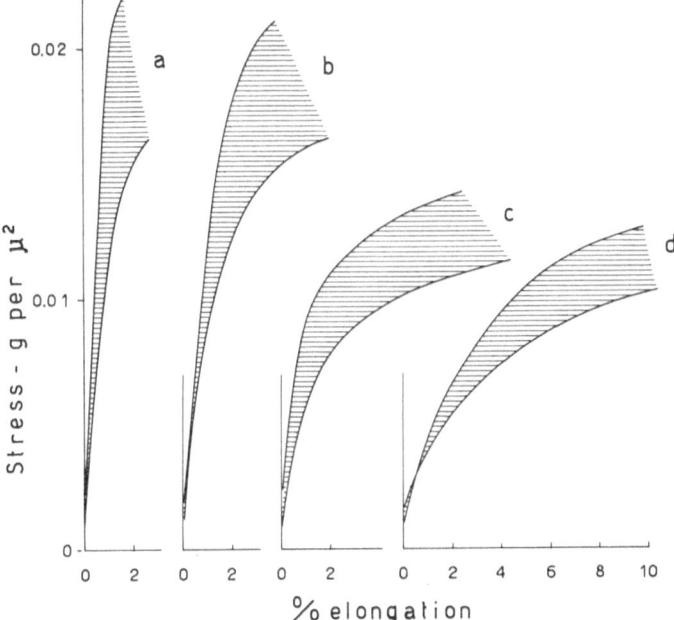

Fig. 5a—d. Diagrams indicating the interval covered by the tensile stress-strain curves recorded for osteons having a marked longitudinal spiral course of the fibre-bundles in successive lamellae, and taken from a man of 30. a Air-dried and fully-calcified osteons. b Air-dried osteons at the initial stage of calcification. c Wet and fully-calcified osteons. d Wet osteons at the initial stage of calcification

The device is then removed and the disc supporting the lower jaw, still attached to the appendage, is reinserted in the cavity. Only at the moment of applying tensile stress are the disc and the lower jaw set free by loosening the screw fixing the lower jaw. In this way they are attached only to the sample. The combined weight of disc and lower jaw was 4 g and the tensile recording was started when the samples had already been elongated by this load.

Results

The charts in Fig. 5 give a unified conspectus of the behaviour of the stress-strain curves recorded from some types of osteon samples kept in different conditions and taken from the cadaver of a 30-year-old man. Here the single diagrams pertaining to the same set of recordings are not reported, only the whole interval covered by the diagrams themselves, regardless of their distribution. In order to allow direct comparisons between the single sets of recorded curves, the tensile strength is expressed in grams per square micron of section (ordinates) and the elongation is indicated as per cent elongation, referred to the original length of the sample (abscissae). Moreover in order to avoid confusion which might arise from superimposing the charts, the origin of the co-ordinate system is transferred along the abscissae axis for each set of diagrams.

Diagram *a* was obtained from 7 stress-strain curves recorded for air-dried osteon samples. The osteons had the maximum amount of calcium salts present and their fibre-bundles had a marked longitudinal spiral course in successive lamellae. The curves approximate a straight line up to fracture, indicating a proportionality of stress and strain in line with Hooke's Law. Only the uppermost points show a very short deviating curve before the breaking of the sample. The modulus of elasticity is very high, $238,600 \pm 71,200$ kg/cm². The ultimate tensile strength is 0.01973 ± 0.00326 g/μ^2, i.e., 19.73 ± 3.26 kg/mm². The percentage of elongation at breaking-point is 2.15 ± 0.55.

Diagram *b* was obtained from 6 tensile stress-strain curves recorded for air-dried osteon samples with the lowest amount of calcium salts microradiographically demonstrable. The fibre-bundles had a marked longitudinal spiral course in successive lamellae. The curves approximate a straight line, as in the case of osteons having the maximum degree of calcification. However the uppermost points show a more deviating curve before the breaking of the bone. The modulus of elasticity is $203,900 \pm 76,400$ kg/cm² and does not reveal significant differences compared with that of fully-calcified osteons. Nor is the ultimate tensile strength (0.01933 ± 0.00288 g/μ^2, i.e., 19.33 ± 2.88 kg/mm²) significantly changed. The percentage of elongation at breaking-point (4.42 ± 1.41) is increased.

Diagram *c* is a plot of the tensile strain values for 8 wet osteons having the maximum amount of calcium salts present and whose fibre-bundles have a marked longitudinal spiral course in successive lamellae. Here the curves show an elastic range like the corresponding dry osteons. But, as the samples elongate more near the breaking-point, the proportionality between stress and strain ends earlier at a proportional limit equal to about half the breaking stress. Beyond this proportional limit, each later increment in stress causes excessive yielding and permanent plastic deformation. In both the straight part of the curves and the deviating plastic region, greater strains are produced than in dry osteons by a given stress. Consequently, the modulus of elasticity ($129,400 \pm 56,200$ kg/cm²) and the ultimate tensile strength (0.01280 ± 0.00135 g/μ^2, i.e., 12.80 ± 1.35 kg/mm²) are lower than in dry bone, whilst the wet samples show a greater percentage elongation (7.36 ± 3.11) than dried osteons.

Diagram *d* shows the interval covered by 8 tensile stress-strain curves recorded for wet osteons having the lowest amount of calcium salts microradiographically demonstrable and revealing a marked longitudinal spiral course of the fibre-bundles in successive lamellae. The diagram for this type of osteon is again similar to that for fully-calcified wet osteons because it deviates strongly from a straight line as the ultimate tensile strength is approached. However, the initial portion also shows a clear deviation from the curve of proportionality indicating that plasticity begins at an early stage. This finding seriously reduces the possibility of calculating the modulus of elasticity or of attributing any precise meaning to it. Its approximate value is $64{,}400 \pm 30{,}400$ kg/cm². The ultimate tensile strength (0.01109 ± 0.00173 g/μ^2, i.e., 11.09 ± 1.73 kg/mm²) is not significantly different from that of wet osteons at the final stage of calcification. The percentage of elongation reaches 9.46 ± 3.36.

The wet osteons were prepared by hydration of the material using saline solution or distilled water. Contrary to the suggestion of CURREY (1959), our previous (ASCENZI and BONUCCI, 1964) and present investigations have revealed that there is no significant difference in tensile properties between osteons rehydrated using saline solution and osteons rehydrated using distilled water.

The charts in Fig. 6 furnish a comparison between a, i.e., diagram c reported in Fig. 5, and the interval covered by 5 tensile stress-strain curves recorded for decalcified wet human osteons originally containing the maximum amount of

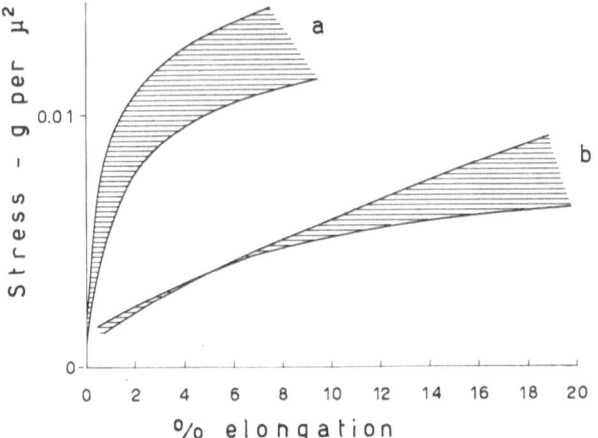

Fig. 6a and b. a The diagram c reported in Fig. 5. b Interval covered from the tensile stress-strain curves recorded from decalcified wet osteons already having the maximum amount of calcium present and having a marked longitudinal spiral course of the fibre-bundles in successive lamellae. Man of 30.

calcium (b). All the samples are characterized by a marked longitudinal spiral course of the fibre-bundles in successive lamellae. Here the modulus of elasticity appears to be very low, equal to $10{,}500 \pm 3{,}500$ kg/cm². The ultimate tensile strength averages 0.00853 ± 0.0014 g/μ^2, i.e., 8.53 ± 1.40 kg/mm². The percentage elongation at the breaking-point is very high, reaching 21.90 ± 7.15.

There is no significant difference between these results and those recorded for decalcified wet osteons having the same longitudinal spiral course of the fibre-bundles in successive lamellae and originally showing the lowest amount of calcium salts, microradiographically demonstrable. In fact, the values measured from 4 osteons are: modulus of elasticity $14{,}800 \pm 6{,}200$ kg/cm²; ultimate tensile strength 0.00877 ± 0.00119 g/μ^2, i.e., 8.77 ± 1.19 kg/mm²; percentage elongation at breaking-point 20.62 ± 5.10.

A comparison between the charts obtained for wet human osteons at both the initial and final stage of calcification, but having fibre-bundles with a marked longitudinal spiral course in successive lamellae and fibre-bundles running alternately under an angle of 90° in successive lamellae, is illustrated in Fig. 7. In respect to diagram a, already given in Fig. 5 (chart c) diagram b, obtained from 5 osteons with the same final stage of calcification, but with fibre-bundles running alternately, also shows in its initial portion a clear deviation from the curve of proportionality, indicating that plasticity begins at an early stage. In this connection the limit of elasticity is not easily determinable, and any attempt to evaluate the modulus of elasticity appears to be unreliable. It is probably something like $64,200 \pm 25,800$ kg/cm². The ultimate tensile strength drops to 0.01074 ± 0.0014 g/μ^2, i.e., 10.74 ± 1.40 kg/mm².

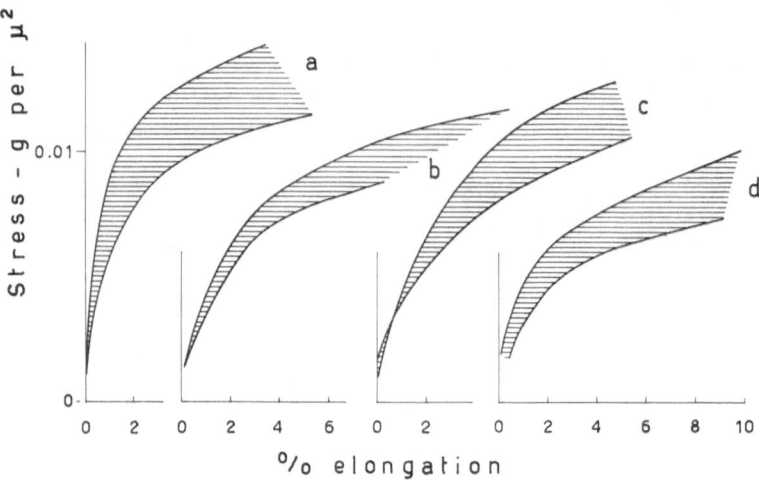

Fig. 7a—d. Diagrams showing the interval covered by the tensile stress-strain curves recorded for osteons taken from a man of 30. a and c Diagrams already reported in Fig. 5 (chart *c* and *d*). b Wet and fully-calcified osteons with fibre-bundles running alternately in such a way that their direction in successive lamellae changes through an angle of 90°. d Wet osteons at initial stage of calcification and with fibre-bundles running alternately in successive lamellae

Results of the same type are furnished by diagram *c*, already reported in Fig. 5 (chart *d*), and diagram *d* obtained from 4 osteons having reached the same initial stage of calcification, but with fibre-bundles running alternately under an angle of 90° in successive lamellae. Here the modulus of elasticity is probably $48,100 \pm 17,200$ kg/cm². The ultimate tensile strength drops to 0.00953 ± 0.00164 g/μ^2, i.e., 9.53 ± 1.64 kg/mm². However, this last result deserves further investigation on a larger number of specimens.

A comparison has been made in order to establish whether any difference exists in the tensile properties of osteons belonging to an adult and to an old man. The measurements concerning the old man were carried out using wet osteons, obtained from a subject aged 80. The samples were prepared from osteons in which the fibre-bundles in one lamella make an angle of nearly 90° with the fibre-bundles in the next one. The osteons having a marked longitudinal spiral course in successive lamellae were discarded because they are very infrequent in bone of old subjects. Diagram *a* (Fig. 8), recorded for 6 osteons in the final stage of calcification and obtained from a man aged 80, reveals no significant differences from diagram *b*, recorded from osteons of the same type and showing the same degree of calcification. Diagram *b* in Fig. 8 corresponds to diagram *b* in Fig. 7. The osteons from the old subject had a modulus of elasticity averaging $86,700 \pm 18,000$ kg/cm² and an ultimate tensile strength averaging 0.00933 $\pm$0.00081 g/μ^2, i.e., 9.33$\pm$0.81 kg/mm².

The results obtained for ox osteons are exactly comparable to those recorded for human osteons. Fig. 9 summarizes the behaviour of the stress-strain curves for both dry and wet osteons having a marked longitudinal spiral course of the fibre-bundles in successive lamellae. Diagram *a* was obtained from 5 curves of air-dried osteon samples with the maximum degree of calcification. The modulus of elasticity averaged 180,700$\pm$19,400 kg/cm² and the ultimate tensile strength was 0.0209$\pm$ 0.00115 g/μ^2, i.e., 20.90$\pm$1.15 kg/mm²; the percentage elongation was 3.45$\pm$0.38.

Diagram *b* is a plot of 5 curves for dried osteons at the initial stage of ossification. The modulus of elasticity is 133,400$\pm$ 24,400 kg/cm²; the ultimate tensile strength 0.01949 $\pm$ 0.00194 g/μ^2, i.e., 19.49 $\pm$1.94 kg/mm²; the percentage elongation 9.32 $\pm$ 1.18.

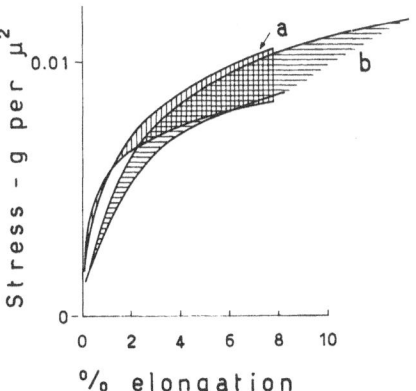

Fig. 8a and b. a Interval covered by the tensile stress-strain curves recorded for wet, fully-calcified osteons with fibre-bundles running alternately in successive lamellae. Man of 80. b This is the same diagram reported in Fig. 7

Diagram *c* is the interval covered by the curves recorded for 5 wet osteon samples with the maximum amount of calcium salts. The modulus of elasticity averages

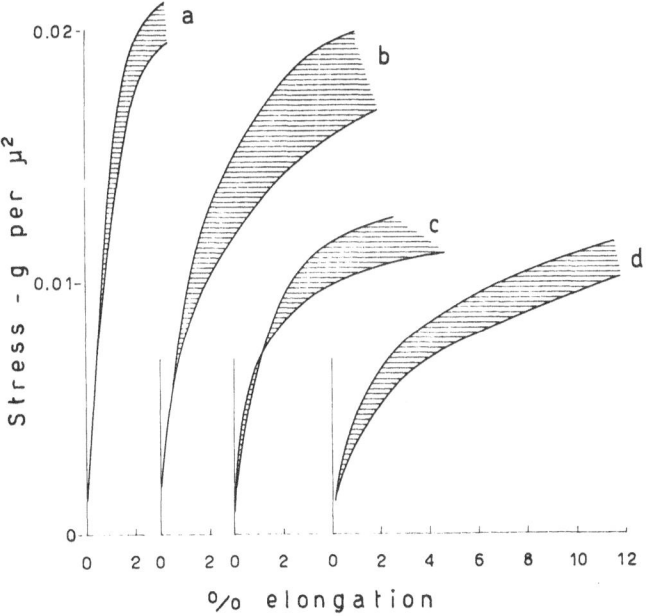

Fig. 9a—d. Diagrams indicating the interval covered by the tensile stress-strain curves recorded for ox osteons having a marked longitudinal spiral course of the fibre-bundles in successive lamellae. a Air-dried and fully-calcified osteons. b Air-dried osteons at the initial stage of calcification. c Wet and fully-calcified osteons. d Wet osteons at the initial stage of calcification

149,600±28,700 kg/cm²; the ultimate tensile strength 0.01212±0.00074 g/μ², i.e., 12.12±0.74 kg/mm²; the percentage elongation 8.30±1.10.

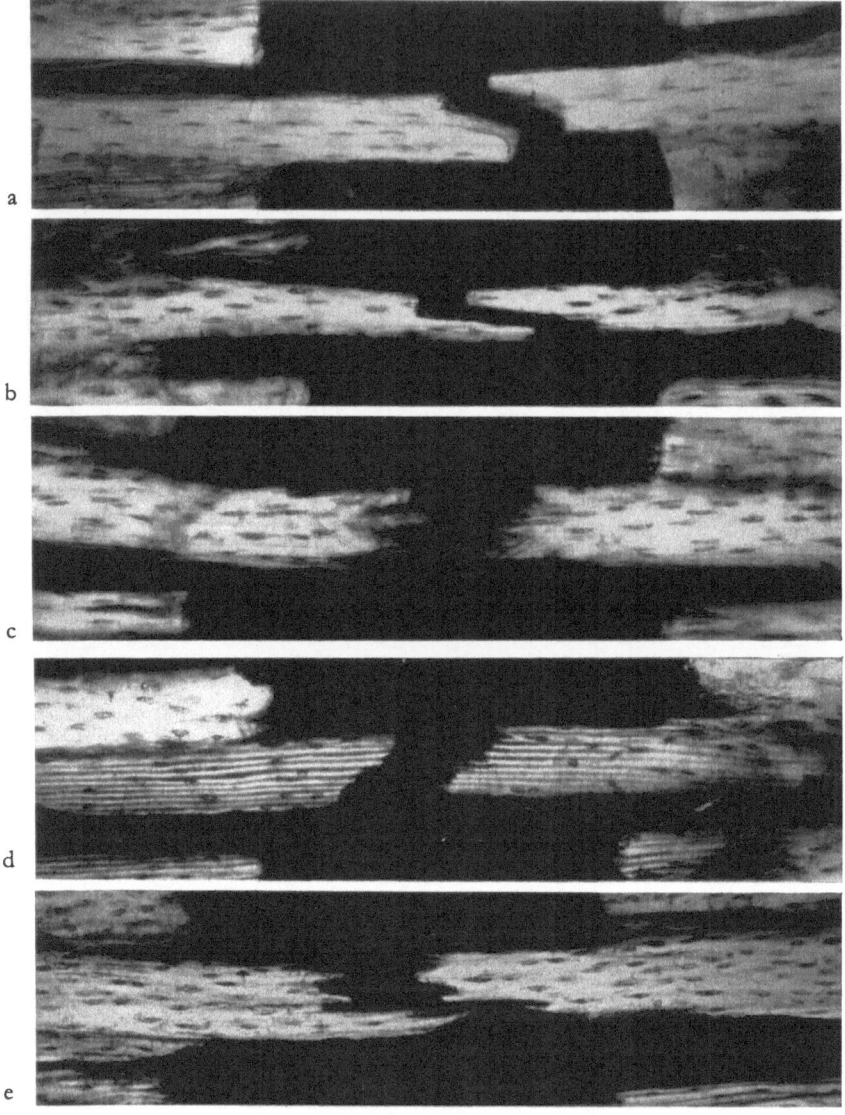

Fig. 10a—e. A series of break ends of tested osteons from a man of 30, under polarizing microscope. a Air-dried fully-calcified osteon with longitudinal course of the fibre-bundles. b Wet fully-calcified osteon with longitudinal course of the fibre-bundles. c Wet osteon at the initial stage of calcification with longitudinal course of the fibre-bundles. d Wet fully-calcified osteon with fibre-bundles changing direction in successive lamellae through an angle of 90°. e Decalcified wet osteon already having the maximum amount of calcium present. ×100

Diagram *d* was obtained from 6 stress-strain curves recorded for wet osteon samples at the initial stage of ossification. The modulus of elasticity is 54,600±

12,300 kg/cm²; the ultimate tensile strength 0.01128 ± 0.00048 g/μ^2, i.e., 11.28 ± 0.48 kg/mm²; the per cent elongation averages 11.14 ± 1.41.

Lastly, a careful examination of the break areas of all the tested osteons was carried out. As seen in Fig. 10 the outlines of the fractures are very variable in shape and in size. They rarely reveal a straight fracture. More often there is considerable irregularity in their shape. The break area may include some osteocyte cavities but there is no demonstrable relationship between the number of intersected osteocytes on the one hand and the crosssection and ultimate tensile strength of the specimen on the other. Besides, the thickness of the specimens with respect to the thinness of the canaliculi prevents a suitable investigation of the role of the same canaliculi in inducing fracture of the osteons.

Discussion

In discussing the results obtained by the present investigation it may be useful to attempt a comparison between the tensile behaviour of single osteons and the tensile behaviour of bone samples of macroscopic size.

In agreement with the results of WERTHEIM (1847), HALLERMANN (1934), MARIQUE (1945), and EVANS and LEBOW (1951), on macroscopic specimens, even in samples taken from isolated osteons having a marked longitudinal arrangement of the fibre-bundles in successive lamellae, the tensile curve for dry bone approximates a straight line. However, in the osteon samples immediately before the breaking-point is approached, the curve deviates very slightly from a straight line. This deviation is somewhat clearer in osteons at the initial stage of calcification than in fully-calcified ones. This result does not allow us to consider dry osteons as entirely elastic as was formerly believed especially by HALLERMANN for macroscopic bone samples. On the other hand, there is some similarity between our data and those reported by DEMPSTER and LIDDICOAT (1952), according to which Hooke's Law is not strictly respected as the breaking-point is approached, the elongation of the sample being somewhat higher than one would expect. In any case, it is necessary to consider that the samples prepared from single osteons are not entirely representative of all the structures present in compact bone. For instance, the interosteonic cementing zone is lacking. Possibly this structure possesses some different mechanical properties according to its high degree of calcification. This view is supported by our findings on osteon calcification, according to which totally calcified units have a straighter stress-strain curve than units at the initial stage of ossification.

There is no significant difference between the modulus of elasticity of dry fully-calcified osteons and the modulus of elasticity of dry osteons at the initial stage of calcification. An impressive similarity exists between our values and those of RAUBER (1876), MARIQUE (1945), EVANS and LEBOW (1951), and DEMPSTER and LIDDICOAT (1952). In this regard Table 1 gives a unified conspectus in which our data are compared with those of the authors cited here.

The influence of moisture on the tensile properties of osteons is indicated by the difference in the shape of stress-strain curve from wet and dry samples. If one considers units having a marked longitudinal arrangement of the fibre-bundles in successive lamellae, the curve shows an elastic range like the dry osteon. But, as

Table 1. *Modulus of elasticity to tension. Comparison between macroscopical and osteonic samples*

Authors	Bone type	Dry $\times 10^6$lbs/in²	kg/cm²	Wet $\times 10^6$lbs/in²	kg/cm²
Rauber (1876)	femur	2.90	203,892		
	tibia	2.67	187,721		
Marique (1945)	femur		200,000 183,962		
Evans and Lebow (1951)	femur	2.67	187,721	2.27	159,597
Dempster and Liddicoat (1952)	femur tibia humerus	2.86 2.69	201,079 189,127	1.77 1.73	124,443 121,632
Ascenzi and Bonucci (1965)	femur		238,600 * 203,900 **		129,400 * 64,400 **

* Osteons at the last stage of calcification and with longitudinal course of the fibre-bundles.

** Osteons at the initial stage of calcification and with longitudinal course of the fibre-bundles.

the samples elongate further toward the breaking-point the proportionality between stress and strain ends earlier at a proportional limit about half the breaking stress. Beyond this proportional limit, each later increment in stress causes excessive yielding and permanent plastic deformation.

The percentage of elongation under tension reveals a significant difference between dry and wet specimens, these last being liable to a greater elongation.

There is a very close correspondence between these results and the curves recorded from macroscopic bone specimens (cf. Dempster and Liddicoat, 1952; Evans and Lebow, 1951). Moreover, our previous (cf. Ascenzi and Bonucci, 1964) and present observations furnish evidence that the ultimate strength of osteons, both at the initial and last stages of calcification, is significantly greater in dry tested samples than in wet ones.

If the values obtained by Evans and Lebow (1951) for the percentage elongation under tension were calculated applying the same procedure used by us, i.e., multiplying by 100 the ratio "initial length/total elongation at breaking-point" one must conclude that the percentage of elongation of the osteons is somewhat greater than that obtained from macroscopic bone samples.

Both dry and wet osteons show that the amount of calcium salts plays an essential role in the elasticity of bone tissue. In dry osteons at the initial stage of ossification the terminal uppermost points of the stress-strain curve deviate much more from the straight line than they do in dry fully-calcified osteons, revealing a greater plastic region. On the other hand, the curve obtained from wet osteons at the initial stage of ossification is similar to that for fully-calcified wet osteons; however, the initial portion shows a clear deviation from the straight line of proportionality indicating that plasticity begins at an early stage. These data clearly indicate that the elasticity of bone tissue is a function of the amount of calcium salts, i.e., of hydroxyapatite crystallites. The lower the calcium content, the higher the plasticity of the tissue (see also CURREY, 1962). In agreement with the present results the stress-strain curves recorded for decalcified osteons indicate a sharp drop in elasticity.

An important point to be considered here is the close similarity between stress-strain curves recorded for human and ox osteons. This finding agrees with the view of ASCENZI and BONUCCI (1961) that, as regards optical properties (birefringence), no essential difference exists between bone units pertaining to two mammal species. Also the investigations of ROWLAND et al. (1959), giving evidence that differences in the amount of calcium salts cannot be demonstrated in human and ox osteons, seem to provide indirect support for this view.

As previously reported the tensile properties of human organic matrix (ossein), prepared both from osteons with the lowest degree of calcification and fully calcified osteons, do not reveal any significant differences. These results appear to be in agreement with those published by STRANDH (1960) who has furnished chemical evidence that variations in the amount of ossein are not demonstrable during the calcification of osteons. Moreover, it must be pointed out that there is a very close correspondence between the modulus of elasticity of decalcified osteons with the lowest and highest degree of calcification. In the former, the modulus of elasticity is $10,500 \pm 3,500$ kg/cm^2 and in the latter $14,800 \pm 6,200$ kg/cm^2. These values are in agreement with those for the modulus of elasticity of collagen which is about 200,000 lb./in.^{2}, i.e., 14,061 kg/cm^2 (cf. CURREY, 1962).

The values given for the ultimate tensile strength of decalcified osteons are not comparable with those reported in our previous paper (ASCENZI and BONUCCI, 1964) because in the present investigation we were able to measure the section of the samples directly on decalcified material.

The differences between the stress-strain curves recorded for osteons having longitudinally-orientated fibres and osteons in which the fibre-bundles run alternately, changing their direction in successive lamellae through an angle of 90^0, suggest that the latter have lower elasticity than the former. Besides, the ultimate tensile strength of osteons in which the fibre-bundles run alternately, changing their direction in successive lamellae, is significantly lower than in osteons having longitudinally-orientated fibres. Consequently, the percentage elongation increases. These results appear to be in agreement with MAJ and TOAJARI's investigations (see MAJ and TOAJARI, 1937a, b; TOAJARI, 1937) revealing that in the specimens with the greatest breaking load, the majority of the collagen fibres are longitudinally orientated. In this connection it should be noted that, according to DEMPSTER and COLEMAN (1961), the cross-grain ultimate tensile strength of bone is much less than the parallel-to-grain strength.

The stress-strain curves of osteon samples from an 80-year-old man show that for wet units whose fibre-bundles change direction in successive lamellae through an angle of 90^0 there is no appreciable change in tensile properties with advancing age. These findings are in agreement with those of EVANS and LEBOW (1951), but differ from those of RAUBER (1876) who reported a decrease in tensile strength of bone with advancing age. In conclusion, our results suggest that the quality of osteons may be unchanged in advanced old age.

As the last topic to be discussed, we want to stress the finding previously mentioned, i.e., that our investigations do not furnish decisive evidence that osteocytes, or their processes, are devoid of any particular significance in inducing fractures in the osteon samples subjected to tests of tensile strength. In fact, it cannot be excluded that these structures have importance from a mechanical point of view. Our investigations are continuing and at this moment definitive conclusions cannot be drawn.

Conclusion and Summary

The results of the present investigation concerning the tensile properties of single osteon specimens and carried out using a purposely devised microwave extensimeter lead to the following conclusions:

1. Drying osteon specimens increases their tensile strength and their modulus of elasticity but reduces their percentage elongation under tension.

2. The osteon samples tested wet have a greater percentage elongation under tension than do dry samples. This is evident from the shape of the stress-strain curve which is almost a straight line to fracture for the dry samples and a curve for wet samples.

3. In the osteon samples tested wet the degree of calcification induces a significant variation of the stress-strain curve indicating an increase in the modulus of elasticity with increasing amounts of calcium salts.

4. The modulus of elasticity to tension of the organic matrix of decalcified osteons corresponds to the modulus of elasticity of collagen.

5. In the osteons having a marked longitudinal arrangement of fibre-bundles in successive lamellae the ultimate tensile strength and modulus of elasticity seem greater and the percentage elongation under tension seems lower than in osteons with fibre-bundles running alternately in such a way that their direction in successive lamellae changes through an angle of about 90⁰.

6. The age of an individual seems to have little influence on the tensile properties of osteons.

7. Human and ox osteons reveal the same tensile behaviour.

8. The comparison of tensile properties in osteon specimens and in macroscopic bone specimens suggests that the osteon is actually the mechanical unit of compact bone.

Acknowledgments

The authors wish to express their indebtedness to G. Ciampi and F. Castellano for technical assistance during the course of the present investigations.

References

Ascenzi, A., and E. Bonucci: A quantitative investigation of the birefringence of the osteon. Acta anat. (Basel) 44, 236—262 (1961).
— — The ultimate tensile strength of single osteons. Acta anat. (Basel) 58, 160—183 (1964).
—, and C. Fabry: Technique for dissection and measurement of refractive index of osteons. J. biophys. biochem. Cytol. 6, 139—142 (1959).
Battaglia, A., F. Bruin, and A. Gozzini: Microwave apparatus for the measurement of the refraction, dispersion and absorption of gases at relatively high pressure. Nuovo Cim. 7, 1—9 (1958).
Carothers, C. O., F. C. Smith, and P. Calabrisi: The elasticity and strength of some long bones of the human body. Naval Med. Res. Inst. Project NM 001 056. 02. 13, 1—18 (1949).

CURREY, J. D.: Differences in the tensile strength of bone of different histological types. J. Anat. (Lond.) **93**, 87—95 (1959).
— Strength of bone. Nature (Lond.) **195**, 513—514 (1962).
DEMPSTER, W. T., and R. F. COLEMAN: Tensile strength of bone along and across the grain. J. appl. Physiol. **16**, 355—360 (1961).
—, and R. T. LIDDICOAT: Compact bone as a non-isotropic material. Amer. J. Anat. **91**, 331—362 (1952).
EVANS, F. G.: Studies in human biomechanics. Ann. N.Y. Acad. Sci. **63**, 586—615 (1955).
— Stress and strain in bones. Their relation to fractures and osteogenesis. Springfield (Ill.): Ch. C. Thomas 1957.
— Relations between the microscopic structure and tensile strength of human bone. Acta anat. (Basel) **35**, 283—301 (1958).
—, and M. LEBOW: Regional differences in some of the physical properties of the human femur. J. appl. Physiol. **3**, 563—572 (1951).
— — The strength of human compact bone as revealed by engineering technics. Amer. J. Surg. **83**, 326—331 (1952).
HALLERMANN: Die Beziehungen der Werkstoffmechanik und Werkstofforschung zur allgemeinen Knochenmechanik. Z. orthop. Chir. **62**, 347—360 (1935).
KOCH, J. C.: The laws of bone architecture. Amer. J. Anat. **21**, 177—298 (1917).
MAJ, G., e E. TOAJARI: Osservazioni istologiche sulle fratture del tessuto osseo normale e dopo decalcificazione. Boll. Soc. ital. Biol. sper. **12**, 57—59 (1937a).
— — La resistenza meccanica del tessuto osseo lamellare compatto misurata in varie direzioni. Boll. Soc. ital. Biol. sper. **12**, 83—86 (1937b).
MARIQUE, P.: Etudes sur le fémur. Anatomie: axes et angles. Déformations. Résistance. Bruxelles: Stoops 1945.
MESSERER, O.: Ueber Elasticität und Festigkeit der menschlichen Knochen. Stuttgart: Cotta 1880.
MONTGOMERY, G.: Technique of microwave measurements. New York: McGraw Hill Book Co. 1947.
RAUBER, A. A.: Elasticität und Festigkeit der Knochen. Leipzig: Engelmann 1876.
ROWLAND, R. E., J. JOWSEY, and J. H. MARSHALL: Microscopic metabolism of calcium in bone. III. Microradiographic measurements of mineral density. Radiat. Res. **10**, 234—242 (1959).
SMITH, J. W., and R. WALMSLEY: Factors affecting the elasticity of bone. J. Anat. (Lond.) **93**, 503—523 (1959).
STRANDH, J.: Microchemical studies on single haversian systems. Exp. Cell Res. **19**, 515—530 (1960).
TOAJARI, E.: Resistenza meccanica ed elasticità del tessuto osseo studiata in rapporto alla minuta struttura. Monit. zool. ital. **48**, 148—154 (1937) Suppl.
WERTHEIM, M. G.: Mémoire sur l'élasticité et la cohésion des principaux tissus du corps humain. Ann. Chim. Phys. (Paris) **21**, 385—414 (1847).

Physical and Histological Differences between Human Fibular and Femoral Compact Bone*

F. G. Evans and S. Bang

Introduction

Although there is a voluminous literature on the histology of bone, its physical properties have been studied far less extensively and there are even fewer attempts to relate the physical properties of bone to its histological structure.

MAJ and TOAJARI (1937) determined the breaking load, under bending, of 12 small specimens of bone cut in a parallel, a radial, and a tangential direction to the long axis of the shaft of ox tibias. They found that the load required to break the bone was 3 times greater for specimens cut in a longitudinal direction than for specimens cut tangentially to the long axis of the bone and 6 times greater than that for specimens cut radially to the long axis of the bone. In addition, the tangentially cut specimens were twice as strong, as far as breaking load was concerned, as the radially cut specimens. The difference in the bending load required to break the specimens was explained on the basis of the predominant direction of collagen fibers in the individual test specimens. In the longitudinally cut specimens, which had the highest breaking load, the predominant direction of the collagen fibers was parallel to the major axis of the specimens. However, in the radially cut specimens (the weakest ones) few of the collagen fibers were parallel to the major axis of the specimens. From the results of their tests, MAJ and TOAJARI concluded (1) that the resistance of compact bone to bending failure is directly proportional to the number of collagen fibers in the plane of the section of the bone; (2) that the cohesiveness of the interfibrillar calcified substance was at least six times inferior to that of the collagen fibers; (3) the anisotropism of bone is the result of the distribution and direction of the collagen fibers; and (4) that the interfibrillar substances probably confer homogeneity and isotropic properties to the compact bone.

OLIVO (1937) studied the mechanical behaviour and microscopic structure of specimens of bone from various areas of metacarpals and metatarsals of the ox, the horse, the dog, and the chamois. Areas with the highest resistance to fracture and a high modulus of elasticity had a predominance of osteones with vertical or steeply spiralling collagen fibers. In those areas with a low resistance to fractures the collagen fibers were more circular and obliquely oriented. However, no actual values for breaking load of the specimens were given.

The relation between breaking load and porosity of bone specimens from the anterior and posterior aspects of a femur of a man 79 years of age was studied

* This research was supported (in part) by Research Grant No. AM 03865-06 from the National Institutes of Health.

by MAJ (1938). The test specimens were taken at 2 cm intervals from strips of bone cut the length of the femur. The porosity was determined by computing the volume of the cavities in the specimens. A similar analysis was made of the metatarsal of an ox. It was found that the breaking load decreased and the porosity increased from the middle of the femur toward each end of the bone. This was particularly true for the breaking load of specimens from the posterior aspect of the femoral shaft. The decrease in the breaking load and the increase in the porosity were not directly proportional. MAJ concluded (1) that the degree of porosity of the specimens was not responsible for variations in breaking load, except in the distal metacarpal range where the porosity was approximately 50 per cent, and (2) that differences in the strength of compact bone in various areas of a skeletal segment are probably the function of intrinsic properties of the bone, e.g., variations in its density and the orientation of its collagen fibers.

The histology of compact bone in relation to its breaking strength and modulus of elasticity was studied by TOAJARI (1938) in bones which he considered to be continuously subjected to a particular type of mechanical stress. Test specimens parallel to the long axis of the bone were prepared from rings of bone from the radius, the ulna and olecranon process of oxen and from the metacarpals and meta-tarsals of oxen, horses, and mules. The specimens were progressively loaded to failure under bending. In addition to the breaking load the deformation of the specimens, in hundredths of a mm per 5 kg load over a span of 7 mm, was also measured.

From his 271 observations TOAJARI concluded that, in general, the modulus of elasticity and the breaking strength of bone are directly proportional and dependent upon the quality and orientation of the collagen fibers. Many differences, as far as its breaking load was concerned, were found in the histological structure of the same skeleton segment. In quite a large percentage of his cases the areas of a bone usually subjected to tension had a higher resistance to fracture and relatively higher modulus of elasticity than areas subjected to compression. Areas with predominantly vertical or steeply spiralling collagen fibers were stronger and had a higher modulus of elasticity than regions with circularly or obliquely oriented collagen fibers.

EVANS (1958) reported that the ultimate tensile strength of standardized specimens of human fibular compact bone is greater, for both wet and dry tested specimens, than that of corresponding specimens of human femoral bone. He found that cross sections of fibular specimens were characterized by having relatively few large osteones and their fragments (remnants) while femoral specimens typically had many small osteones and fragments. It was suggested that a few large osteones and fragments make the ultimate tensile strength of a given amount of bone greater than do a large number of small osteones and fragments. This was related to the abundance of cementing lines because they represent areas of weakness or lower resistance to tensile failure. Thus, the larger the number of osteones and fragments the greater the number of cementing lines and hence the weaker the ultimate tensile strength of the specimens. Polarized light studies of the specimens indicated that the predominant direction of the collagen fibers was also a possible factor, the greater the number of collagen fibers parallel to one another and to the long axis of the specimen and of the intact bone the higher its ultimate tensile strength. Because of the relatively small number of sections studied the author pointed out that "some of the present conclusions were tentative and may have to be modified by results of subsequent studies."

The relation of the tensile strength of compact bone from the shaft of ox femurs to different histological types of bone was studied by Currey (1959). A strong negative correlation was found between the amount of reconstruction that recurred in a piece of bone, and hence the number of Haversian systems in it, and its tensile strength. Two complementary explanations for the findings were given: (1) immature Haversian systems have a large central cavity which reduces the actual amount of bone substance present per unit volume and (2) newly formed Haversian systems are not fully mineralized and therefore presumably weaker than the surrounding bone.

Recently Ascenzi and Bonucci (1964) found that the ultimate tensile strength of single osteones was greatest in those osteones having the majority of the collagen fibers longitudinally oriented, i.e., parallel to the long axis of the test specimen.

The present investigation, based on additional material, is an extension of the previous one (Evans, 1958) but involving the relation between more physical properties of human cortical bone and its histological structure. The physical properties involved are ultimate tensile strength (stress) and strain, single shearing strength, modulus of elasticity, hardness, density, and the ratio between the single shearing strength and ultimate tensile strength. The histological structure of the specimens is studied under ordinary and polarized light and the distribution of calcium by means of microradiography.

The following report deals with the relation of secondary osteones, remnants or fragments of secondary osteones, interstitial lamellae and spaces to the above, previously determined, physical properties of the specimens.

Materials and Methods

The physical properties mentioned above were determined for specimens of cortical bone which were machined to a standardized size and shape from the proximal, the middle, and the distal thirds of the shaft of femurs and fibulas of adult, embalmed dissecting room cadavers. Each third of the femoral shaft was also subdivided into anterior, posterior, medial, and lateral quadrants but the fibula was too small to obtain test specimens from quadrants of the shaft.

The shape and dimensions of the tensile and shearing test specimens as well as the methods of determining their ultimate tensile and shearing strength, tensile strain, modulus of elasticity, and hardness were the same as those used in earlier studies by Evans and Lebow (1951).

Whenever a force or load is applied to a body, e.g., a test specimen of a material, it tends to deform the body. The tendency of a body to be deformed by the application of a force is resisted by an intermolecular force which tends to restore the body to its pretest dimensions and form. The *ratio* of this internal or "restoring" force to the area upon which the force is assumed to act is *stress* or strength, as it is often called. Because stress is a ratio, i.e., force per unit area, it can never be seen, no matter how large it is. Stress is always a derived quantity and can only be computed in terms of load or force per unit area (kg/mm^2 or lbs/in^2).

Strain is the change in the linear dimensions of an object as the result of the application of a force. Strain has no standard units of measurement and can be measured in terms of inches/inch, mm/mm, or percentage. If the strain is sufficiently large it can be seen, e.g., the stretching of a rubber band.

During the tensile tests the force (load) was uniformly applied over the cross-sectional area of the specimen so it was quite easy to compute the ultimate tensile strength (stress) of the specimen from the formula $S = \dfrac{P}{A}$ in which S equals stress (strength), P equals force or load and A equals the cross-sectional area of the specimen. The tensile tests were made by slowly applying the load in the direction of the long axis of the specimen which coincided with the long axis of the intact bone. The speed of loading was constant throughout the tests.

In the shearing tests the load was applied to the specimen in a direction perpendicular to its long axis and that of the intact bone. The magnitude of the shearing stress (strength) can be computed from the formula $S_s = \dfrac{P}{A_s}$ in which S_s is shearing stress, P is force or load and A_s is the cross-sectional area of the specimen where the shearing occurs. The force was parallel to the cross-sectional area of the specimen upon which the stress is computed and was projected through the center of the area.

The tensile stress was plotted against the tensile strain to obtain a stress-strain curve. From a tangent to the straightest part of the stress-strain curve the *tangent modulus* of elasticity, or the ratio of unit stress to unit strain, was computed. The *modulus of elasticity* is a measure of the *stiffness* of a material.

The hardness of the specimens was determined with a Rockwell Superficial Hardness Tester, as described by Evans and Lebow (1951). The tensile and shearing strength, modulus of elasticity, tensile strain, and hardness were determined from wet or moist specimens because they more nearly approximate the condition of bone in the living body.

The density of the specimens was determined by means of a Strontium[90] densitometer developed by Evans, Coolbaugh, and Lebow (1951). Air dried specimens were used to avoid the moisture trapped within the spaces of the specimens when they were wet. In dry specimens the spaces were filled with air which would provide little, if any, obstruction to the passage of the beta rays.

After the above physical properties of the specimens had been determined, an analysis of variance was made on IBM 7090 computers in The University of Michigan Computer Center to determine if there were statistically significant differences between the mean values for the various physical properties of the fibular and femoral specimens. Ninety-five percent confidence intervals were also computed for some cases. The programing for the statistical analysis was done in consultation with Miss Esther Schaeffer, a member of the staff of The University of Michigan Statistical Laboratory.

Cross sections of the specimens were taken as near as possible to the fracture site and prepared for histological study. Two kinds of sections were studied: (1) decalcified sections 20 μ thick, embedded in celloidin, and (2) undecalcified ground sections 40—75 μ thick. Photographs of the section, in ordinary and polarized light, were taken with a Zeiss photomicroscope and then analyzed by the photomicrograph-weight method employed by Evans (1958). The method is based on the use of photomicrographs of cross sections of bone specimens enlarged to a standardized size on photographic paper of a constant known weight. The weight of the photographic paper (in grams) was determined from the average weight of 10 pieces of paper each 100 cm² in area.

On the basis of the weight of 100 cm² of the photographic paper a conversion factor (K), for converting paper weight (in grams) into area (mm² or μ^2), was computed from the formula $K = \frac{10^4}{pF^2}$ in which F equals the linear enlargement of the positive print and p equals the weight of 100 cm² of the photographic paper. By cutting out and weighing various histological elements visible in the enlarged photomicrograph of a section it was possible to determine the area (mm² or μ^2) of each of the different elements and the percentage of the total section formed by each of them.

The first step in the analysis of a section was the determination of the number of osteones and their fragments (remnants). The enlarged photomicrograph was then cut out, weighed, and its area (mm²) determined. This area was called the *original break area*. It was not, however, identical with the break area, as measured with calipers after testing, because the caliper measurements did not allow for irregularities in the margins of the specimen.

The paper representing Haversian canals, and resorption spaces in the enlarged photomicrograph was then carefully cut out, weighed and its area determined. The remainder of the enlarged photomicrograph of the section was then weighed and its area determined. This area was called the *corrected break area* and, except for the area of the lacunae and the canaliculi, represented the area of the bone tissue actually subjected to stress during a test.

The same procedure was used in determining the following information with respect to the corrected break area: (1) area of the osteones, (2) area of the fragments or remnants of osteones, (3) area of interstitial lamellae, (4) the percent of the corrected break area formed by osteones, (5) the percent of the corrected break area formed by fragments (remnants) of osteones, (6) the percent of the corrected break area formed by interstitial lamellae, (7) the number of osteones/mm², (8) the number of fragments of osteones/mm², (9) the average area (μ^2) per osteon, and (10) the average area (μ^2) per fragment of osteon. Sections from 54 femoral and 37 fibular specimens randomly selected have been analyzed in this way. The femoral specimens were obtained from 6 individuals and the fibular ones from 8 individuals.

The histological data thus obtained was subjected to an analysis of variance on computers to determine if there were statistically significant differences between the means for the various histological elements.

Coefficients of correlation were calculated on the computers to determine the degree of correlation (1) between the various physical properties of the bone specimens, (2) between the different histological elements of the specimens, and (3) between physical properties of the specimens and their histological elements.

Results

The results of the physical property determinations and of the histological study of cross sections of the test specimens at the fracture site are present below in tables and in graphs. Differences between the means for the physical properties and for the various histological components of the cross sections were considered statistically significant only if they were at the 0.01 or 0.02 significance level. The same was true for the coefficients of correlation between the physical properties and the histological components of the cross sections.

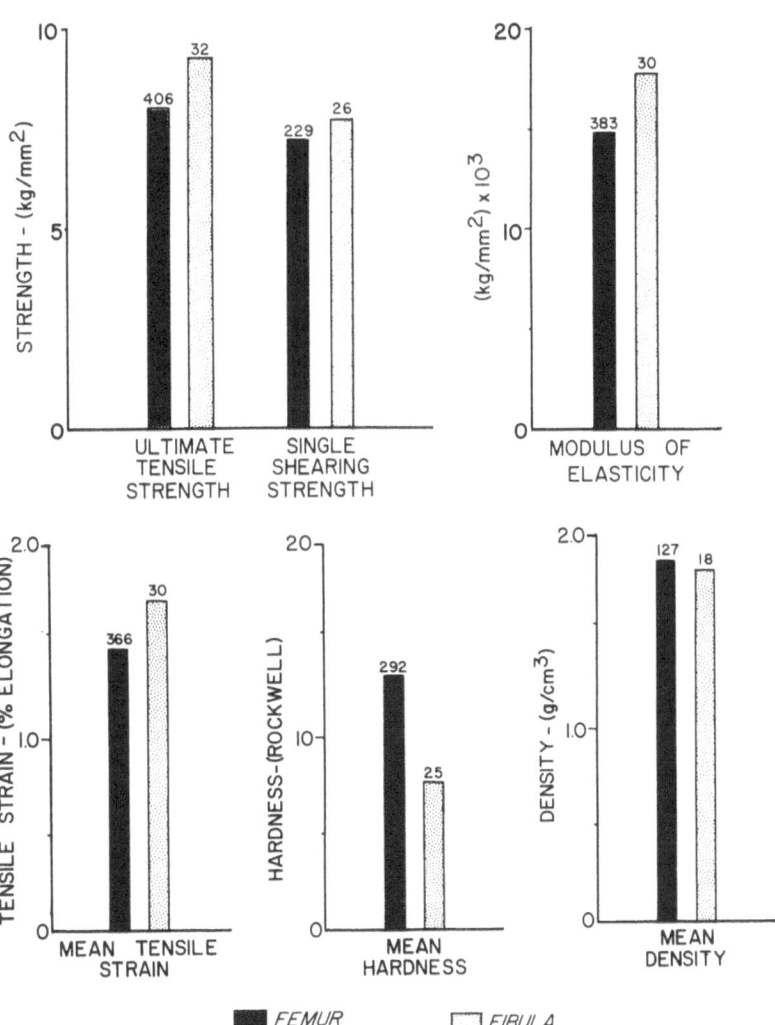

Fig. 1. Mean for physical properties of standardized specimens of cortical bone from the shaft of adult embalmed human femora and fibulae. The number of specimens is indicated at the top of each bar

Comparison of the means for the physical properties of the femoral and the fibular specimens (Fig. 1) showed that the fibular specimens were stronger (ultimate tensile and single shearing strength), stiffer (modulus of elasticity) and had a greater tensile strain (% elongation) than the femoral specimens. The latter were harder (Rockwell number) and a little denser than the fibular specimens.

An analysis of variation between the means for the physical properties (Table 1) revealed a highly significant difference between the means for the modulus of elasticity and for the ultimate tensile strength of the femoral and the fibular specimens. The means for the other physical properties of the two types of specimens were not significantly different.

Table 1. *Variation beetwen physical properties of embalmed adult human femoral and fibular cortical bone. Physical property data obtained from wet specimens unless otherwise indicated*

Variable	N (1)	N (2)	F-Ratio	Mean (1)	±Sigma (1)	Mean (2)	±Sigma (2)
Modulus of elasticity (kg/mm^2)	383	30	20.869 **	1476.87	490.87	1777.89	310.01
Ultimate tensile strength (kg/mm^2)	405	32	15.606 **	7.98	2.01	9.44	1.93
Tensile strain (% elongation)	366	30	3.462	1.45683	0.66417	1.71367	1.27521
Single shearing strength (kg/mm^2)	299	26	4.635 *	7.20	1.07	7.70	1.55
Hardness (Rockwell number)	292	25	2.554	13.199	17.207	7.600	10.943
Density (dry specimens) (g/cm^3)	127	18	5.543 *	1.87512	0.08159	1.82111	0.14295
$\dfrac{\text{SSS}}{\text{UTS}} \times 100$	229	26	1.061	87.2022	11.7722	84.4615	20.2229

  * Between 5% and 1% significance.
  ** 1% or better significance level.
  (1) = Femur, (2) = Fibula.
  Significant differences: Modulus of elasticity — fibula greater than femur. Ultimate tensile strength — fibula greater than femur.

Highly significant differences (Table 2) were also found between the histological components of cross sections of the femoral and the fibular specimens. The femoral specimens had more osteones/mm^2, a greater percentage of the corrected break area (CBA) formed by osteones, and a greater percentage of the original break area (OBA) formed by spaces than did the fibular specimens. The latter had a greater percentage of the corrected break area formed by interstitial lamellae (Figs. 2 and 3). The differences between the other histological components of the cross sections were not statistically significant.

A high degree of correlation between some of the physical properties and histological elements of the two types of sections was likewise found (Table 3). The femoral specimens exhibited a strong positive correlation, at the 0.02 significance level, between tensile strain (% elongation) and the percentage of the corrected break area (CBA) formed by osteones. An even stronger negative correlation, at the 0.01 significance level, was found between the modulus of elasticity (E)

HISTOLOGICAL STRUCTURE

ORIGINAL BREAK AREA

HUMAN FEMUR

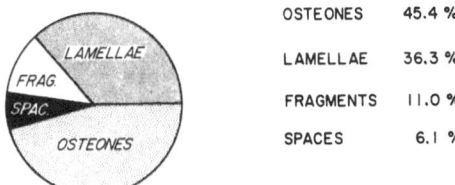

OSTEONES	45.4 %
LAMELLAE	36.3 %
FRAGMENTS	11.0 %
SPACES	6.1 %

HUMAN FIBULA

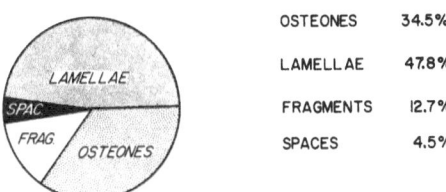

OSTEONES	34.5%
LAMELLAE	47.8%
FRAGMENTS	12.7%
SPACES	4.5%

Fig. 2. The mean percentage of the original break area of 54 femoral and 37 fibular specimens, based on cross sections of the specimens at the level of the fracture site, formed by various histological elements. The lacunae and the canaliculi are not included in the spaces. See text for definition of original break area

HISTOLOGICAL STRUCTURE

CORRECTED BREAK AREA

HUMAN
FEMUR

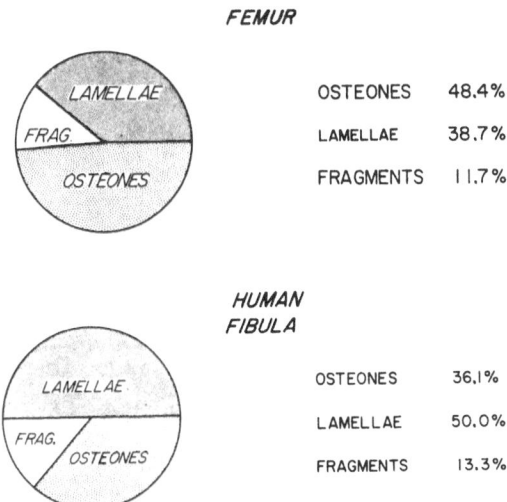

OSTEONES	48.4%
LAMELLAE	38.7%
FRAGMENTS	11.7%

HUMAN
FIBULA

OSTEONES	36.1%
LAMELLAE	50.0%
FRAGMENTS	13.3%

Fig. 3. The mean percentage of the corrected break area of 54 femoral and 37 fibular specimens, based on cross sections of the specimens at the level of the fracture site, formed by various histological elements. See text for definition of corrected break area

Table 2. *Variation between the means for histological elements in 54 femoral and 37 fibular specimens of adult human cortical bone*

Variable	Femur mean	Fibula mean	F-ratio	Significance level (%)
Number of osteones/mm^2	12.87	8.69	36.967	0.01
Number of osteon fragments/mm^2	3.72	3.88	0.178	--
Cross-sectional area/osteon (μ^2)	39.2	43.5	3.168	—
Cross-sectional area/fragment of osteon (μ^2)	33.3	36.9	2.066	—
% CBA formed by osteones	48.44	36.12	28.232	0.01
% CBA formed by osteon fragments	11.71	13.30	1.713	—
% CBA formed by interstitial lamellae	38.72	50.03	15.031	0.01
% OBA formed by spaces	6.05	4.46	7.664	0.01

CBA = Corrected Break Area, OBA = Original Break Area.

Table 3. *Correlation between physical properties and histological structures of embalmed adult human cortical bone*

Variable	Femur			Fibula		
	N	R	Significance level	N	R	Significance level
E vs % OBA, spaces	42	−0.441	0.01	36	−0.171	—
UTS vs % OBA, spaces	43	−0.600	0.01	39	−0.293	--
% Elongation vs osteones/mm^2	41	−0.013	—	36	0.421	0.01
% Elongation vs % CBA, osteones	41	0.361	0.02	36	0.493	0.01
% Elongation vs % CBA, lamellae	41	−0.212	---	36	−0.476	0.01
% Elongation vs % OBA, spaces	41	−0.313	—	36	−0.384	0.02
SSS vs area/osteon fragment	43	−0.039	—	38	−0.484	0.01
Hardness vs % OBA, spaces	43	−0.533	0.01	36	−0.208	—

CBA = Corrected Break Area; E = Tangent Modulus of Elasticity; N = Number of Specimens; OBA = Original Break Area; R = Correlation Coefficient; SSS = Single shearing Strength (Stress); UTS = Ultimate Tensile Strength (Stress).

and the percentage of the original break area (OBA) formed by spaces, between the ultimate tensile strength (UTS) and the percentage of the original break area formed by spaces, and between hardness and the percentage of the original break area (OBA) formed by spaces.

The fibular specimens had a higher positive correlation, at the 0.01 significance level, between tensile strain and the percentage of the corrected break area formed by osteones than did the femoral specimens. In contrast to the femoral specimens they had a high positive correlation, at the same significance level, between tensile strain and the number of osteones/mm². They differed from the femoral specimens in having a strong negative correlation between tensile strain and the percentage of the corrected break area formed by interstitial lamellae, between single shearing strength (SSS) and the area/osteon fragment (remnant) and between tensile strain and the percentage of the original break area formed by spaces.

Discussion

Because all the test specimens used in the present study were obtained from embalmed bones the question arises as to what effect embalming may have had on the physical properties of the bone. In order to answer this question an analysis of variation between the means for the same physical properties was made for 193 embalmed wet tested and 209 unembalmed wet tested specimens of cortical bone from adult human tibias. The analysis revealed that the mean hardness and the mean density of the embalmed specimens was significantly greater (1% or better significance level) than the means for the same physical properties of the unembalmed specimens. No significant differences were found between the means for the other physical properties of the embalmed and unembalmed specimens.

An interesting result of the study was the differences found in the histological composition of the cross sections of the femoral and the fibular specimens. The mean number of osteones/mm², the percentage of the corrected break area formed by osteones, and the percentage of the original break area formed by spaces was significantly greater in the femoral than in the fibular sections. The latter had a significantly greater percentage of the corrected break area formed by interstitial lamellae. Thus, the femoral sections were characterized by having many small osteones and relatively little interstitial lamellae while the fibular sections had relatively few, but larger, osteones and more interstitial lamellae. The percentage of spaces in the original break area of the femoral sections was greater than that in the original break area of the fibular sections. The significantly greater percentage of spaces (holes) in the original break area of the femoral specimens as measured with calipers immediately after a test, would tend to weaken them as compared with the fibular specimens, because the spaces act as stress raisers or sites of higher stress concentration. However, the propagation of a microfracture in the femoral specimens might be less than in the fibular ones because the fracture would stop when it reached a hole or space.

These combinations of histological components suggest that osteones tend to reduce the tensile strength and increase the tensile strain of cortical bone. Interstitial lamellae, on the other hand, tend to increase the tensile strength of cortical bone although, at least in fibular bone, they seem to reduce its tensile strain.

The negative relation between tensile strength and the relative number of osteones in a given area of cortical bone was noted previously by EVANS (1958) for human bone and by CURREY (1959) for bovine bone. Relations between other physical properties of cortical bone and its histological structure were not investigated.

The differences in the histological structure of the femoral and the fibular sections account, in part, for the significantly greater ultimate tensile strength and modulus of elasticity of the fibular specimens as compared with the femoral ones. The probable reason for these significant differences is not the number of osteones, osteon fragments and interstitial lamellae in themselves but as previously suggested by Evans (1958), the differences in the amount of cement lines or cementing substance in the femoral and the fibular specimens. The actual amount of cementing substance in a section could not be measured directly, but the relative amount of it could be indirectly determined from the number of osteones and osteon fragments in a section because these histological elements are bounded by cementing lines. Thus, the greater the number of osteones and osteon fragments in a section the greater the amount of cementing lines or substance.

Cement lines bound osteones and osteon fragments and are important, as far as the strength characteristics and the modulus of elasticity of bone are concerned, because they consist of mucopolysaccharide (Eastoe, 1956) and are not crossed by collagen fibers from adjacent osteones or fragments. Consequently, they represent areas of weakness where failure, stretching, or slipping can occur between adjacent osteones, osteon fragments, and interstitial lamellae.

That cement lines actually do represent areas of weakness where failure will occur has been noted by Maj and Toajari (1937) who reported, for bone specimens from ox tibiae, that the fracture lines usually circled around osteones instead of going between or across osteon lamellae. Dempster and Coleman (1961), studying human bone, also found ". . . that rupture lines tend to follow the curvature of the cement lines surrounding Haversian systems. Often, in a transverse section, a fracture will follow the curvature of the cement line of a Haversian system and then veer across interstitial bone and beyond until it reaches the cement line of an adjacent Haversian system. Less commonly, fracture lines tend to break across a few Haversian lamellae to follow a curved course between lamellae of the system for a short distance before breaking away." They concluded, on the basis of photomicrographs of fracture regions in wet and dry tested undecalcified specimens, that "weaker structural elements are the cement lines surrounding osteones and the planes between the lamellae of Haversian systems . . ."

Fractures produced in transverse sections of decalcified specimens by applying tension in the plane of the section showed a tendency to follow the cement lines indicating that they were sites of weakness (Fig. 4). A similar phenomenon was seen in a section that had been allowed to dry (Fig. 5). During drying tensile stresses and strains are created in the bone as the result of which the bone shrinks and separations occur along the cement lines.

Another interesting result of the study, although it did not show the desired significance level of 0.01 or 0.02, was the larger mean size (cross-sectional area) of the fibular osteones as compared with the femoral osteones. A similar relation between size of osteon and size of bone is seen in a table, given by Aoji (1959), showing the mean cross-sectional area of osteones in the middle of the shaft of human femora, tibiae, fibulae, humeri, radii, and ulnae. His data showed a progressive increase in osteon size from the femur to the tibia to the fibula and from the humerus to the radius to the ulna. In other words, the smaller the bone the larger the osteones.

The other physical properties of bone studied in the present investigation (hardness and density) showed no correlations with osteones, fragments, and lamellae.

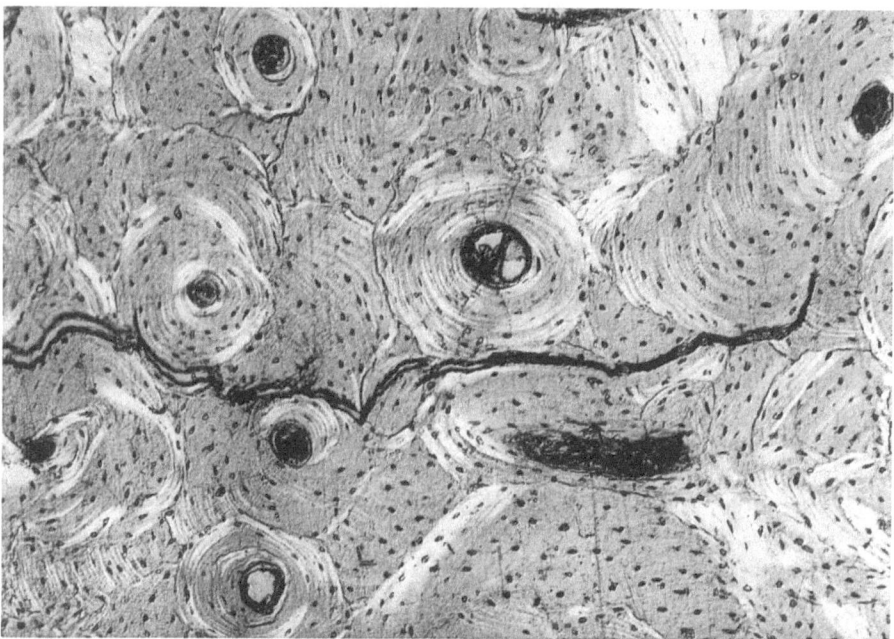

Fig. 4. Fracture produced in a cross section of a decalcified femoral specimen by tension in the plan of the section. Note the tendency of the fracture to follow the cement lines

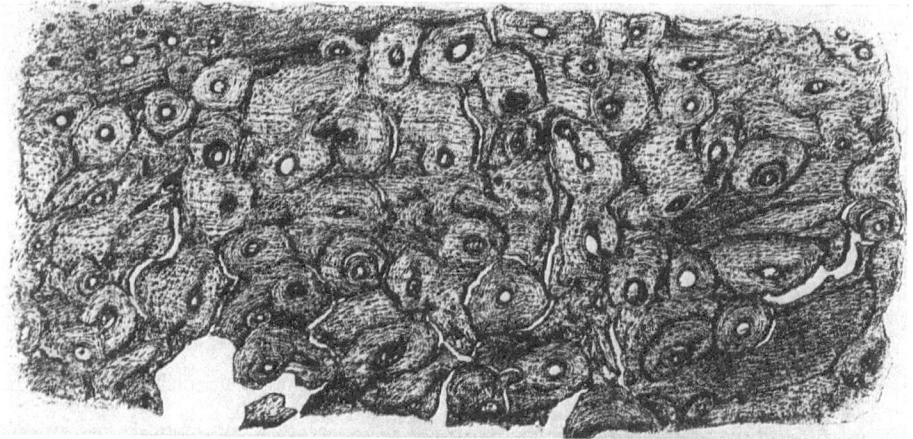

Fig. 5. Fractures produced by drying of a cross section of a delcalcified femoral specimen. Shrinkage during drying created tensile stresses in the specimen which resulted in tensile failures tending to follow the cementing lines

However, they are probably influenced by the amount and distribution of calcium in the section, a problem now under investigation in our laboratory. An intensive study is now being made of the relation of the orientation and distribution of collagen fibers and of calcium to the physical properties of specimens from different bones.

It should be remembered that the preceding discussion of the histological structure of bone as related to its physical properties was based on transverse sections of the specimens at or as near the fracture site as it was possible to make them. The relative proportion of the cross-sectional area formed by spaces, by osteones, by fragments of osteones, and by interstitial lamellae at or near the fracture site would be somewhat different at other levels of the specimen because, as shown by Fila-gamo (1946), Koltze (1951), and Cohen and Harris (1958), osteones are branching structures rather than simple tubes. However, the histological structure seen at the fracture site indicates that it was the weakest arrangement of the various components; otherwise the fracture would not have occurred where it did.

Summary and Conclusions

1. The ultimate tensile and single shearing strength, tensile strain (% elongation), tangent modulus of elasticity, hardness (Rockwell number), density, and ratio of single shearing strength to ultimate tensile strength were determined for 45 femoral and 32 fibular specimens of embalmed human cortical bone. Specimens of a standardized size and shape were used. Density was determined for air dried specimens but all other physical properties were determined from wet or moist specimens.

2. The tensile specimens were cut and loaded in the direction of the long axis of the specimen and of the intact bone. The shearing tests were made in a direction perpendicular to the long axis of the specimen and the intact bone.

3. The strength tests were made under static loading in a materials testing machine calibrated to an accuracy of $\pm 1\%$.

4. An analysis of variance of the physical properties revealed that the mean ultimate tensile strength and the mean modulus of elasticity of the fibular specimens were significantly greater (1% or better significance level) than the means for the corresponding physical properties of the femoral specimens. There were no significant differences between the means for the other physical properties.

5. Analysis of variation between the means for various histological components of femoral and fibular cortical bone revealed that the number of osteones/mm², the percentage of the corrected break area (area of section minus the spaces) formed by osteones, and the percentage of the original break area formed by spaces was significantly greater (1% or better significance level) in the femoral than in the fibular sections. The latter had a significantly greater percentage of the corrected break area formed by interstitial lamellae.

6. Correlation coefficients showed a strong positive correlation between tensile strain (% elongation) and the percentage of the corrected break area formed by osteones for both the femoral and the fibular specimens. The fibular specimens also had a strong positive correlation between tensile strain and the number of osteones/mm². Equally strong negative correlations were found between the percentage of the original break area formed by spaces and the modulus of elasticity, the ultimate tensile strength, and the hardness of the femoral specimens. The fibular specimens exhibited a strong negative correlation between tensile strain and the percentage of the corrected break area formed by interstitial lamellae, between

tensile strain and the percentage of the original break area formed by spaces, and between single shearing strength and the cross-sectional area/osteon fragment (remnant).

7. Cement lines represent areas of weakness in bone where failure, slipping, or stretching can occur. Thus, the greater the number of osteones and osteon fragments in a given area of bone the larger the amount of cementing substance and the lower the tensile strength and modulus of elasticity of the bone.

8. Interstitial lamellae appear to increase the tensile strength of bone. The greater the proportion they form of a given area of bone the higher its tensile strength.

9. Femoral bone is characterized by having many small osteones and their fragments and relatively little interstitial lamellae in contrast to fibular bone which has relatively few large osteones and fragments but more interstitial lamellae. Therefore, femoral bone has more cementing substance for a given area than does fibular bone which explains, in part, the significantly lower tensile strength and modulus of elasticity of femoral bone in comparison with fibular bone.

10. The relation of collagen fiber orientation and calcium distribution to the physical properties of cortical bone is also being investigated.

11. Whether or not other bones exhibit similar relations between physical properties and histological structure as found in our femoral and fibular specimens remains to be seen.

References

AOJI, O.: Metrical studies on the lamellar structure of human long bones. J. Kyoto Prefect. Med. Univ. **65**, 941—965 (1959) [Japanese with English summary].

ASCENZI, A., and E. BONUCCI: The ultimate tensile strength of single osteones. Acta anat. (Basel) **58**, 160—183 (1964).

COHEN, J., and W. H. HARRIS: The three-dimensional anatomy of Haversian systems. J. Bone Jt Surg. A **40**, 419—434 (1958).

CURREY, J. D.: Differences in the tensile strength of bone of different histological types. J. Anat. (Lond.) **93**, 87—95 (1959).

DEMPSTER, W. T., and R. F. COLEMAN: Tensile strength of bone along and across the grain. J. appl. Physiol. **16**, 355—360 (1961).

EASTOE, J. E.: The organic matrix of bone. In: The biochemistry and physiology of bone, p. 81—105 [G. H. BOURNE (ed.)]. New York: Academic Press 1956.

EVANS, F. G.: Relations between the microscopic structure and tensile strength of human bone. Acta anat. (Basel) **35**, 285—301 (1958).

— C. C. COOLBAUGH, and M. LEBOW: An apparatus for determining bone density by means of radioactive strontium (Sr^{90}). Science **114**, 182—185 (1951).

—, and M. LEBOW: Regional differences in some physical properties of the human femur. J. appl. Physiol. **3**, 563—572 (1951).

FILOGAMO, G.: La forme et la taille des ostéones chez quelques mammifères. Arch. Biol. (Liège) **57**, 137—143 (1946).

KOLTZE, H.: Studie zur äußeren Form der Osteone. Z. Anat. Entwickl.-Gesch. **115**, 584—596 (1951).

MAJ, G.: Osservazioni sulle differenze topografiche della resistenza meccanica del tessuto osseo di uno stesso segmento schelectrico. Monit. zool. ital. **49**, 139—149 (1938).

—, e E. TOAJARI: Osservazioni sperimentali sul meccanismo di resistenze del tessuto osseo lamellare compatto alle azioni meccaniche. Chir. Organi Mov. **22**, 541—557 (1937).

OLIVO, O. M.: Rispondenza della funzione meccanica varia degli osteoni con la loro diversa minuta architettura. Boll. Soc. ital. Biol. Sper. **12**, 400—401 (1937).

TOAJARI, E.: Resistenza meccanica ed elasticità del tessuto osseo studiata in rapporto alla minuta struttura. Monit. zool. ital. **48** (1938).